Advances in Measurements and Instrumentation: Reviews

Book Series, Volume 1

S. Y.Yurish
Editor

Advances in Measurements and Instrumentation: Reviews

Book Series, Volume 1

International Frequency Sensor Association Publishing

S. Y. Yurish, *Editor*
Advances in Measurements and Instrumentation: Reviews, Book Series, Vol. 1

Published by IFSA Publishing, S. L., 2018
E-mail (for print book orders and customer service enquires):
ifsa.books@sensorsportal.com

Visit our Home Page on http://www.sensorsportal.com

Neither the authors nor IFSA Publishing accept any responsibility or liability for loss or damage occasioned to any person or property through using the material, instructions, methods or ideas contained herein, or acting or refraining from acting as a result of such use.

ISBN: 978-84-09-07321-4
e-ISBN: 978-84-09-07322-1
BN-20181215-XX
BIC: TBM
BISAC: TEC022000

Contents

Preface

It is my great pleasure to present the first volume of the *Advances in Measurements and Instrumentation: Reviews* book Series started by the IFSA Publishing in 2018.

The book volume contains seven chapters written by nine contributors from academia and industry from 6 countries: Algeria, Canada, China, Germany, Slovak Republic and United Kingdom.

Chapter 1 describes methods and algorithms, based on using the classical regression approach used for the errors-in-variable model and applied it for the comparative calibration with the generalized polynomial calibration function. The present approach is based on strictly metrologist considerations. The inputs quantities are fully characterized by their state-of-knowledge distributions, derived from the associated uncertainty budgets of the direct measurements from the calibration experiment, which allows to combine type A and type B methods of evaluation of the measurements, as well as to incorporate correlated measurements with the comparative calibration, i.e. with the situation when both variables entering the calibration experiment are subject to errors (including the used measurement standard).

Chapter 2 aims at giving a brief, yet mostly complete instructive overview about the fundamental basics of sampling theory and its application in terms of Fourier transform and other in their continuous and discrete version. The chapter is split into three successive sections, which point at different levels of abstraction and complexity.

Chapter 3 presents mathematical tools for measurements for quality control based material testing and characterization applications.

Chapter 4 discuss water cut measurement technologies, methods discussed and identifies their gaps.

Chapter 5 introduces three popular methods to model the colored noise. The methods include the shaping filter based power spectral density, auto-regressive moving average based on time series, and Allan variance. The three methods are provided in detail with verification by examples.

Chapter 6 describes recent developments in open channel cross section design. Based on a review of 115 publications, this chapter focuses on recent developments related to all factors, except longitudinal slope, which is beyond the scope of this review.

Chapter 7 reviews recent hydrologic Muskingum models for channel flood routing, which are particularly useful at the initial planning stage where the gauging system is still underdeveloped or insufficient for precise measurements. This chapter is organized as follows. Section 2 presents a historical perspective of the development of hydrologic Muskingum models, including the motivation for the surge of recent research. Section 3 reviews original Muskingum models (1959-2014). Section 4 reviews two new Muskingum models (2013-2014) that have started an alternative thinking to improve model performance. Section 5 reviews 15 subsequently developed models (2014-present). Section 6 describes common solution methods of calibration. Section 7 discusses practical considerations and Section 8 presents concluding remarks.

I hope that readers will enjoy this book and it will be a valuable tool for those who are involved in research and development in appropriate area.

Sergey Y. Yurish

Editor
IFSA Publishing   Barcelona, Spain

Contributors

Salah Bouhouche
Research Center in Industrial Technologies CRTI, P.O.Box 64, Cheraga 16014, Algiers, Algeria

Said M. Easa
Department of Civil Engineering, Ryerson University, Toronto, Ontario, Canada M5B 2K3, E-mail: seasa@ryerson.ca

Matthias Ratajczak
Institute of Fluid Dynamics, Helmholtz-Zentrum Dresden-Rossendorf, Dresden, Germany, E-mail: m.ratajczak@hzdr.de

Martin Seilmayer
Institute of Fluid Dynamics, Helmholtz-Zentrum Dresden-Rossendorf, Dresden, Germany, E-mail: m.seilmayer@hzdr.de

Prafull Sharma
Corrosion Radar Ltd., Cambridge, United Kingdom

Kedong Wang
School of Astronautics, Beihang University, Beijing 100191, China

Gejza Wimmer
Faculty of Science, Matej Bel University, Banská Bystrica, Slovak Republic and Mathematical Institute, Slovak Academy of Sciences, Štefánikova 49, 814 73 Bratislava, Slovak Republic, E-mail: wimmer@mat.savba.sk

Viktor Witkovský
Institute of Measurement Science, Slovak Academy of Sciences, Dúbravská cesta 9, 841 04 Bratislava, Slovak Republic, E-mail: witkovsky@savba.sk

Hoi Yeung
Cranfield University, Bedford, United Kingdom

13

Chapter 1

Generalized Polynomial Comparative Calibration: Parameter Estimation and Applications

Viktor Witkovský and Gejza Wimmer

1.1. Introduction

According to the International vocabulary of metrology (VIM) [1], *calibration is defined as an operation that, under specified conditions, in a first step, establishes a relation between the quantity values with measurement uncertainties provided by measurement standards and corresponding indications with associated measurement uncertainties and, in a second step, uses this information to establish a relation for obtaining a measurement result from an indication.*

A calibration may be expressed by a statement, calibration function, calibration diagram, calibration curve, or calibration table. Often, the first step alone in the above definition is perceived as being calibration. According to VIM, *calibration curve* is expression of the relation between indication and corresponding measured quantity value. It is understood that the calibration curve expresses a one-to-one relation between the indication and the quantity value, at least for a specified range of the considered quantity values.

This chapter deals with the comparative calibration, i.e. with the situation when both variables entering the calibration experiment are subject to errors (including the used measurement standard).

Viktor Witkovský
Institute of Measurement Science and Mathematical Institute of the Slovak Academy of Sciences, Bratislava, Slovak Republic

Here we assume that the calibration function expresses the ideal (true, errorless) values of the measurand in scale (units) of the measuring instrument Y (typically the less precise measuring instrument, the calibrated device) as a function of the values of the measurand in scale of the measuring instrument X (typically the more precise instrument, the standard). In other words, the calibration function expresses the relationship between the errorless values of the measured quantities expressed in units of the used measuring instruments, X and Y, respectively. In particular, we are interested in finding the proper point and interval estimators of the coefficients of well-defined calibration function.

Here, the considered calibration function is assumed to be a *polynomial of given degree p* or a general nonlinear function, here related to the polynomial of a given degree p. In most practical situations, however, the comparative calibration is based on a simple *linear calibration function.*

In fact, the polynomial calibration function is linear in its parameters but due to the errors in both variables the considered calibration model becomes nonlinear in the model parameters. As we shall present below, after proper linearization of the given nonlinear model, the considered polynomial calibration model can be represented (approximately) as a linear errors-in-variables (EIV) model.

From statistical point of view, our goal is to suggest a procedure for fitting the considered generalized polynomial calibration function and deriving the approximate confidence intervals or regions for the parameters of this function.

The present approach to the calibration problem, however, is based on a metrological approach to the expression of uncertainty in measurement, [2] and [3]. It combines the current state-of-knowledge, expressed by the probability distributions about values attributed to the considered measurands by using the measurement standard and the calibrated device, and the statistical regression techniques based on using the EIV model with restrictions on the model parameters.

Here we discuss the methods for solving the problem, including the iterative algorithm for estimating the parameters of the calibration function and specification of their state-of-knowledge distributions, and deriving the associated coverage intervals.

Moreover, the procedure for measuring by using the calibrated device is described. Under given restrictions the proposed interval estimators are precise and fully satisfactory for practical use. Application of this comparative calibration model and the suggested estimation procedure is further illustrated by *calibration of industrial platinum resistance thermometer*.

Novelty of the considered approach to the polynomial comparative calibration rests on its ability to incorporate correlated data and to combine the type A as well as the type B evaluations of the measurements. For more details and metrological interpretation see the international standards on evaluation of measurement data, the *Guide to the expression of uncertainty in measurement* (GUM) [2] and its *Supplements* [3], [4], and/or the broad follow-up discussions in the metrology and measurement science literature.

In order to simplify the used mathematical expressions, throughout the chapter we shall use the standard vector-matrix notation. In particular, A' or a' denotes the transpose of a matrix A or a vector a, a^k denotes element-wise kth powers of the elements of a vector a, A^{-1} denotes inverse of a full-ranked square matrix A, $A = Diag(a)$ denotes a diagonal matrix with elements on its diagonal created by the elements of a vector a, and $A = [B \vdots C]$ denote a matrix A concatenated by sub-matrices (or column vectors) of suitable dimensions, B and C. Moreover, $\mu \sim X$ means that μ is distributed as X.

1.2. Measurement Procedure

Throughout the chapter we shall consider the generalized polynomial calibration model and assume that the following assumptions about the model and the measurement process hold true:

Derivation of the calibration function (i.e. specification of the best estimates of the parameters, their state-of-knowledge distributions and/or their associated coverage intervals) is based on realization of a calibration experiment which allows to specify (by using and combining type A and type B methods of evaluation) the state-of-knowledge distributions about quantity values of the considered measurands expressed in scales of both instruments, X (the more precise one) and Y (the less precise one).

The measurands (quantities intended to be measured) are represented here by a set of I suitably chosen objects (state of the phenomenon, body, or substance carrying the quantity), say $V_1, ..., V_I$, such that their true values μ_i, $i = 1, ..., I$, expressed in scale of the instrument X, span the pre-planned calibration range (that of instrument X).

For the more precise instrument X (the standard) we shall assume that the observed measurement result x_i is a realization of the random variable X_i, i.e. $x_i = X_i^{(obs)}$, modelled by the measurement equation

$$X_i = (\mu_i + Z_{X,i}) + (\delta_{X,i} + \Delta_{X,i}) + \nabla_X, \qquad (1.1)$$

for all considered objects (measurands) $i = 1, ..., I$, where

- μ_i are the true (unknown) values of considered quantities of interest in units of the more precise measuring device X, for all $i = 1, ..., I$.

- $Z_{X,i}$ are independent random variables representing our knowledge about the measurement errors, with known zero-mean distributions (typically normal or t-distribution) and given standard deviations $u_{Z_{X,i}}$, $i = 1, ..., I$, obtained from type A evaluations.

- $\delta_{X,i}$ are the known applied corrections (scalar values) of the systematic effects, obtained from type B evaluations for all $i = 1, ..., I$. Without loss of generality, here we shall assume that $\delta_{X,i} = 0$.

- $\Delta_{X,i}$ are independent random variables representing our knowledge about the considered individual corrections, with known zero-mean distribution and given standard deviations $u_{\Delta_{X,i}}$, $i = 1, ..., I$, obtained from type B evaluations.

- ∇_X, is a random variable representing our knowledge about the applied correction common to all measurements realized with the measuring device X, represented by independent random variable with known zero-mean distributions with standard uncertainties u_{∇_X}, obtained from type B evaluations.

All considered corrections $\Delta_{X,i}$ and ∇_X, as well as the measurement errors $Z_{X,i}$ are assumed to be mutually independent random variables.

Hence, based on the measurement equation (1.1) and the observed value x_i of X_i, we can specify the (unrestricted) state-of-knowledge distribution about the possible values of μ_i (i.e. distribution of the values attributed to the measured quantity in scale of the measuring device X), given as the distribution of the random variable $\tilde{\mu}_i$, for all $i = 1, \dots, I$,

$$\tilde{\mu}_i \sim \left(x_i - Z_{X,i}\right) - \left(\delta_{X,i} + \Delta_{X,i}\right) -$$

$$- \nabla_X \sim \left(x_i - \delta_{X,i}\right) - \left(Z_{X,i} + \Delta_{X,i} + \nabla_X\right) \sim \tilde{x}_i - \tilde{X}_i \qquad (1.2)$$

Here, $\tilde{x}_i$ represents the *best estimate* of μ_i (nonstochastic scalar value, corrected for the systematic effects) and the random variable $\tilde{X}_i$ represents our knowledge about the distribution (shifted to zero-mean) of the values attributed to the measured quantity. For more details see [2].

Similarly, for the less precise instrument Y (the calibrated device) we shall assume that the observed measurement result y_i is a realization of the random variable Y_i, i.e. $y_i = Y_i^{(obs)}$, modelled by the measurement equation

$$Y_i = (v_i + Z_{Y,i}) + (\delta_{Y,i} + \Delta_{Y,i}) + \nabla_Y \qquad (1.3)$$

for all considered objects $i = 1, \dots, I$, where

- v_i are the true (unknown) values of considered quantities of interest in units of the less precise (calibrated) measuring device Y, for all $i = 1, \dots, I$.

- $Z_{Y,i}$ are the independent random variables representing our knowledge about the measurement errors, with known zero-mean distributions (typically normal or t-distribution) and given standard deviations $u_{Z_{Y,i}}$, $i = 1, \dots, I$, obtained from type A evaluations.

- $\delta_{Y,i}$ are the known applied corrections (scalar values) of the systematic effects, obtained from type B evaluations for all $i = 1, \dots, I$. Without loss of generality, here we shall assume that $\delta_{Y,i} = 0$.

- $\Delta_{Y,i}$ are the independent random variables representing our knowledge about the considered individual corrections, with known zero-mean distribution and given standard deviations $u_{\Delta_{Y,i}}$, $i = 1, \ldots, I$, obtained from type B evaluations.

- ∇_Y, is a random variable representing our knowledge about the applied correction common to all measurements realized with the measuring device Y, represented by independent random variable with known zero-mean distribution with standard uncertainty u_{∇_Y}, obtained from type B evaluations.

All considered corrections $\Delta_{Y,i}$ and ∇_Y, as well as the measuremet errorss $Z_{Y,i}$ are assumed to be mutually independent random variables.

Based on the measurement equation (1.3) and the observed value y_i of Y_i, we can also specify the (unrestricted) state-of-knowledge distribution about the possible values of v_i (i.e. distribution of the values attributed to the measured quantity in scale of the measuring device Y.), given as the distribution of the random variable $\tilde{v}_i$, for all $i = 1, \ldots, I$,

$$\tilde{v}_i \sim (y_i - Z_{Y,i}) - (\delta_{Y,i} + \Delta_{Y,i}) - \nabla_Y \sim (y_i - \delta_{Y,i}) -$$
$$-(Z_{Y,i} + \Delta_{Y,i} + \nabla_Y) \sim \tilde{y}_i - \tilde{Y}_i \qquad (1.4)$$

In order to simplify the measurement equations (1.1) and (1.3) and the subsequent mathematical formulae, we shall use the vector-matrix notation. For that, let

$$X = (X_1, \ldots, X_I)', \mu = (\mu_1, \ldots, \mu_I)'$$

$$Z_X = (Z_{X,1}, \ldots, Z_{X,I})', u_{Z_X}^2 = \left(u_{Z_{X,1}}^2, \ldots, u_{Z_{X,I}}^2\right)'$$

$$\delta_X = (\delta_{X,1}, \ldots, \delta_{X,I})', \Delta_X = (\Delta_{X,1}, \ldots, \Delta_{X,I})'$$

$$u_{\Delta_X}^2 = \left(u_{\Delta_{X,1}}^2, \ldots, u_{\Delta_{X,I}}^2\right)', Y = (Y_1, \ldots, Y_I)', v = (v_1, \ldots, v_I)'$$

$$Z_Y = (Z_{Y,1}, \ldots, Z_{Y,I})', u_{Z_Y}^2 = \left(u_{Z_{Y,1}}^2, \ldots, u_{Z_{Y,I}}^2\right)'$$

$$\delta_Y = (\delta_{Y,1}, \ldots, \delta_{Y,I})', \Delta_Y = (\Delta_{Y,1}, \ldots, \Delta_{Y,I})'$$

$$u_{\Delta_Y}^2 = \left(u_{\Delta_{Y,1}}^2, \ldots, u_{\Delta_{Y,I}}^2\right)' \qquad (1.5)$$

denote the respective I-dimensional column vectors, and moreover, let $1 = (1, \ldots, 1)'$ denotes the I-dimensional column vector of ones. For simplicity and without loss of generality we shall assume further that $\delta_X = \delta_Y = 0$, where 0 denotes vector of zeros. Then the vector representation of the measurement model for X and Y is given as

$$X = \mu + Z_X + \Delta_X + \nabla_X 1, \quad Y = \nu + Z_Y + \Delta_Y + \nabla_Y 1 \quad (1.6)$$

Based on the above specifications and given assumptions, we get the following expressions for the expected value and the covariance matrix of the joint $(2 \times I)$-dimensional measurement vector $(X', Y')'$,

$$E \begin{pmatrix} X \\ Y \end{pmatrix} = \begin{pmatrix} \mu \\ \nu \end{pmatrix}$$

$$Cov \begin{pmatrix} X \\ Y \end{pmatrix} = \begin{pmatrix} Cov(X) & Cov(X,Y) \\ Cov(Y,X) & Cov(Y) \end{pmatrix} = \begin{pmatrix} \Sigma_X & \Sigma_{XY} \\ \Sigma_{YX} & \Sigma_Y \end{pmatrix} = \Sigma \quad (1.7)$$

with

$$\Sigma_X = \Lambda_X + u_{\nabla_X}^2 E, \quad \Sigma_Y = \Lambda_Y + u_{\nabla_Y}^2 E, \quad \Sigma_{XY} = \Sigma_{YX} = 0, \quad (1.8)$$

where

$$\Lambda_X = Diag\left(u_{Z_X}^2 + u_{\Delta_X}^2\right), \quad \Lambda_Y = Diag\left(u_{Z_Y}^2 + u_{\Delta_Y}^2\right) \quad (1.9)$$

are $(I \times I)$-dimensional diagonal matrices with specified known variances on their diagonals, E denotes $(I \times I)$-dimensional matrix of ones, and 0 $(I \times I)$-dimensional matrix of zeros. Generalization of the measurement model to the situation with $Cov(X,Y) \neq 0$ is possible, but not considered here.

1.3. Calibration Model

We shall consider the calibration model specified as a measurement model (1.6)-(1.9) with a set of possibly nonlinear system of restrictions, which specify the considered calibration function.

1.3.1. Polynomial Calibration Function

As already stated, the calibration function expresses the ideal (errorless) values of the quantities in scale (units) of the measuring instrument Y (the less precise measuring instrument) as a function of the values of the quantities in scale (units) of the measuring instrument X (the more precise instrument), i.e.

$$v = f(\mu), \qquad (1.10)$$

which can be alternatively specified by the set of restrictions

$$F(\mu, v) = 0, \qquad (1.11)$$

where $F(\mu, v) = f(\mu) - v$. Moreover, the calibration function may depend on other parameters, say the vector parameter a. In particular, the *polynomial calibration function* of the order p can be expressed as

$$v = a_0 1 + a_1\mu + \cdots + a_p\mu^p = [1, \mu, \cdots, \mu^p] \begin{pmatrix} a_0 \\ a_1 \\ \vdots \\ a_p \end{pmatrix} = Ma \quad (1.12)$$

thus, by using the vector-matrix notation we get the restriction in the form

$$Ma - v = 0, \qquad (1.13)$$

where v is the I-dimensional vector of the (unknown) errorless quantity values, expressed in scale of the measuring instrument Y, $M = [1, \mu, \cdots, \mu^p]$ is the $(I \times (p+1))$-dimensional matrix with columns given as powers of the (unknown) quantity values μ, expressed in scale of the measuring instrument X, and $a = (a_0, \ldots, a_p)'$ denotes the $(p+1)$-dimensional vector of the (unknown) polynomial parameters.

In most practical and frequent situations only the simple linear calibration function is considered, i.e. the special case of the polynomial calibration function with order $p = 1$. In such a case we get the *linear calibration function* specified by the restrictions

$$a_0 1 + a_1\mu - v = 0 \qquad (1.14)$$

alternatively expressed as

$$Ma - v = 0, \qquad (1.15)$$

where $M = [1, \mu]$ and $a = (a_0, a_1)'$.

On the other hand, the polynomial calibration function (1.12) can be generalized to more complicated calibration functions. In general, we can consider the calibration function specified by the set of restrictions

$$F(\mu, v, a) = 0 \qquad (1.16)$$

Frequently, $F(\mu, v, a) = f(\mu, a) - v$, and $f(\cdot, \cdot)$ denotes a reasonable I-dimensional vector function which depends on the (unknown) quantity values μ and the $(p + 1)$-dimensional vector of the (unknown) parameters, say $a = (a_0, \dots, a_p)'$.

As we shall illustrate below, see the example illustrating calibration of the industrial platinum resistance thermometer, the calibration function can be modelled by the *generalized polynomial calibration function*, with the following specification of the calibration curve:

$$F(\mu, v, a) = Ma - v = 0, \qquad (1.17)$$

where $a = (a_0, \dots, a_p)'$ denotes the $(p + 1)$-dimensional vector of the (unknown) polynomial parameters and M denotes the $(I \times (p + 1))$-dimensional full ranked matrix, specified by its elements $\{M\}_{ij}$, which can be modelled as polynomials of μ_i, i.e.

$$\{M\}_{ij} = \sum_{k=0}^{q} A_{ij,k} \, \mu_i^k, \qquad (1.18)$$

where $A_{ij,k}$ denote the given (known) coefficients for $i = 1, \dots, I$, $j = 1, \dots, p + 1$, and $k = 0, \dots, q$.

1.3.2. Linearized Calibration Model

The calibration model (1.6)-(1.9) with the nonlinear restrictions (1.16) (which is in specific cases equivalent with (1.13), (1.15), or (1.17)) is a nonlinear model in the parameters μ, v, a. In fact, it is a linear regression model with nonlinear restrictions on its parameters.

In order to simplify the numerical approach and the subsequent statistical inference, we shall approximate the model by linearizing the nonlinear system of restrictions, $F(\mu, v, a) = 0$.

By using the first-order Taylor expansion about μ_0, v_0, and a_0 (the approximate prior values of the parameters) we get

$$F(\mu, v, a) \approx F(\mu_0, v_0, a_0) + \frac{\partial F(\mu, v, a)}{\partial \mu'} \Big|_0 (\mu - \mu_0) +$$

$$+ \frac{\partial F(\mu, v, a)}{\partial v'} \Big|_0 (v - v_0) + \frac{\partial F(\mu, v, a)}{\partial a'} \Big|_0 (a - a_0) =$$

$$= A_0 \begin{pmatrix} \mu_\Delta \\ v_\Delta \end{pmatrix} + B_0 a_\Delta + c_0 \qquad (1.19)$$

where the used symbol $\big|_0$ specifies that the presented partial derivatives are evaluated at the prior values μ_0, v_0, and a_0.

In general, we shall use the following notation for the transformed parameters:

$$\mu_\Delta = \mu - \mu_0 \, , v_\Delta = v - v_0, a_\Delta = a - a_0 \qquad (1.20)$$

and the following notation for the matrices and vectors specifying the system of linear restrictions,

$$A_0 = \left[\frac{\partial F(\mu, v, a)}{\partial \mu'} \Big|_0 \; \vdots \; \frac{\partial F(\mu, v, a)}{\partial v'} \Big|_0 \right], B_0 = \frac{\partial F(\mu, v, a)}{\partial a'} \Big|_0 \, ,$$

$$c_0 = F(\mu_0, v_0, a_0) \qquad (1.21)$$

In particular, if the considered *calibration function is a polynomial of the order p*, we get $F(\mu, v, a) = Ma - v = [1, \mu, \cdots, \mu^p]a - v$, as specified by the restrictions (1.12)-(1.13), and hence, based on (1.19) we get

$$A_0 = \left[\frac{\partial Ma}{\partial \mu'} \Big|_0 \; \vdots \; -I \right] = [D_0 \; \vdots \; -I] =$$

$$= \left[Diag\left(a_{0,1} 1 + 2a_{0,2}\mu_0 + \cdots + pa_{0,p}\mu_0^{p-1}\right) \; \vdots \; -I \right]$$

$$B_0 = \frac{\partial Ma}{\partial a'} \Big|_0 = M_0 = [1, \mu_0, \cdots, \mu_0^p]$$

$$c_0 = M_0 a_0 - v_0 = a_{0,0} 1 + a_{0,1}\mu_0 + \cdots + a_{0,p}\mu_0^p - v_0, \quad (1.22)$$

where by I we denote the $(I \times I)$-dimensional diagonal identity matrix.

The simplest situation is for the linear calibration function, i.e. the special case of the *polynomial calibration function with order* $p = 1$. In this case we get the following matrices and vectors,

$$A_0 = [\, D_0 \,\vdots - I\,] = [\, a_{0,1}I \,\vdots - I\,]$$

$$B_0 = M_0 = [1, \, \mu_0]$$

$$c_0 = M_0 a_0 - \nu_0 = a_{0,0}1 + a_{0,1}\mu_0 - \nu_0 \qquad (1.23)$$

On the other hand, the most complicated calibration function considered in this chapter is the *generalized polynomial calibration function*. For the generalized polynomial calibration function specified by (1.17) and (1.18) we get

$$A_0 = \left[\frac{\partial Ma}{\partial \mu'} \Big|_0 \,\vdots - I\right] = [\, D_0 \,\vdots - I\,]$$

$$B_0 = \frac{\partial Ma}{\partial a'} \Big|_0 = M_0$$

$$c_0 = M_0 a_0 - \nu_0 \qquad (1.24)$$

with $\{M_0\}_{ij} = \sum_{k=0}^{q} A_{ij,k}\, \mu_{0,i}^k.$

$$\frac{\partial Ma}{\partial \mu'} \Big|_0 = Diag\left(\frac{\partial \{Ma\}_1}{\partial \mu_1} \Big|_0, \dots, \frac{\partial \{Ma\}_I}{\partial \mu_I} \Big|_0\right) \qquad (1.25)$$

In Section 1.7, we present a real example, leading to such calibration function, together with derivation of the associated matrices A_0, B_0, and the vector c_0.

Now, the (approximate) linear calibration model can be expressed as a linear regression model, specified by the expected values and the covariance matrix of the transformed measurements X_Δ and Y_Δ, with the linear restrictions on the transformed parameters $\mu_\Delta, \nu_\Delta, a_\Delta$, for more details see [5]. In particular,

$$\begin{pmatrix} X_\Delta \\ Y_\Delta \end{pmatrix} \sim \left(\begin{pmatrix} \mu_\Delta \\ \nu_\Delta \end{pmatrix}, \Sigma\right), \quad A_0 \begin{pmatrix} \mu_\Delta \\ \nu_\Delta \end{pmatrix} + B_0 a_\Delta + c_0 = 0, \qquad (1.26)$$

where $X_\Delta = X - \mu_0$, $Y_\Delta = Y - \nu_0$, $\mu_\Delta = \mu - \mu_0$, $\nu_\Delta = \nu - \nu_0$, and $a_\Delta = a - a_0$, and the model covariance matrix Σ is specified by the equations (1.7)-(1.9).

1.4. Estimation of the Model Parameters

The model (1.26) serves as a first-order approximation to the originally nonlinear calibration model, which is correct in the vicinity of the pre-selected fixed values of the parameters, μ_0, ν_0, and a_0.

Based on that, the (μ_0, ν_0, a_0)-locally best linear unbiased estimator (BLUE) of the model parameters μ_Δ, ν_Δ, a_Δ, and its covariance matrix can be estimated. In general, the suggested locally best estimator has the required optimality properties at the vicinity of the selected values μ_0, ν_0, and a_0, only.

1.4.1. Best Linear Unbiased Estimator of the Calibration Model Parameters

In particular, here we shall consider BLUE of the parameters in *the incomplete direct-measurements model with linear constraints of type II*, as suggested in [5, pp. 146-152].

Let us briefly summarize the known results: Let $Z = (Z_1, ..., Z_m)'$ be m-dimensional random vector representing the measurements, with its expected value $E(Z) = \gamma$, where γ is m-dimensional vector of parameters – subject of direct measurements, and with the known covariance matrix Σ. Further, we shall assume that there is another q-dimensional vector of parameters, say θ, which is, however, not a subject of direct measurements, and such that the parameters γ and θ fulfill the linear restrictions $A\gamma + B\theta + c = 0$, specified by a known vector c and given matrices A and B. Thus, the model is specified as

$$Z \sim (\gamma, \Sigma) \ \& \ A\gamma + B\theta + c = 0 \qquad (1.27)$$

According to [5], the best linear unbiased estimators $\hat{\gamma}$ and $\hat{\theta}$ of the vector parameters γ and θ are given as a solution to the following optimization problem

$$\begin{pmatrix} \hat{\gamma} \\ \hat{\theta} \end{pmatrix} = \underset{\gamma, \theta}{\mathrm{argmin}} \left((Z - \gamma)' \Sigma^{-1} (Z - \gamma) - 2\lambda'(A\gamma + B\theta + c) \right), (1.28)$$

where λ is the associated vector of *Lagrange multipliers*. From that we get the following system of linear equations which should be solved (by using the generalized least squares method) to get the BLUEs $\hat{\gamma}$ and $\hat{\theta}$,

$$\begin{pmatrix} A\Sigma A' & B \\ B' & 0 \end{pmatrix}\begin{pmatrix} \lambda \\ \hat{\theta} \end{pmatrix} = \begin{pmatrix} -AZ - c \\ 0 \end{pmatrix} \ \& \ \hat{\gamma} = Z + A'\lambda \qquad (1.29)$$

After simple manipulation, the solution can be alternatively expressed as

$$\begin{pmatrix} \hat{\gamma} \\ \hat{\theta} \end{pmatrix} = -\begin{pmatrix} \Sigma A'Q_{11} \\ Q_{21} \end{pmatrix} c + \begin{pmatrix} I - \Sigma A'Q_{11}A \\ -Q_{21}A \end{pmatrix} Z, \qquad (1.30)$$

where Q_{11} and Q_{12} are the blocks of the matrix Q defined by

$$Q = \begin{pmatrix} Q_{11} & Q_{12} \\ Q_{21} & Q_{22} \end{pmatrix} = \begin{pmatrix} A\Sigma A' & B \\ B' & 0 \end{pmatrix}^{-1} \qquad (1.31)$$

assuming that the required matrix inversion exists, which leads to the following representation of the block matrices

$$Q_{11} = (A\Sigma A')^{-1} - (A\Sigma A')^{-1}B\,(B'(A\Sigma A')^{-1}B)^{-1}B'(A\Sigma A')^{-1}$$

$$Q_{12} = (A\Sigma A')^{-1}B(B'(A\Sigma A')^{-1}B)^{-1}$$

$$Q_{21} = (B'(A\Sigma A')^{-1}B)^{-1}\,B'(A\Sigma A')^{-1}$$

$$Q_{22} = -(B'(A\Sigma A')^{-1}B)^{-1} \qquad (1.32)$$

The covariance matrix of the joint estimator $(\hat{\gamma}', \hat{\theta}')'$ is

$$Cov\begin{pmatrix} \hat{\gamma} \\ \hat{\theta} \end{pmatrix} = \begin{pmatrix} \Sigma - \Sigma A'Q_{11}A\Sigma & -\Sigma A'Q_{12} \\ -Q_{21}A\Sigma & -Q_{22} \end{pmatrix} \qquad (1.33)$$

Now, we apply the BLUE (1.30) and its covariance matrix (1.33) for the linearized calibration model (1.26) with restriction matrices specified in (1.21), resp. (1.22), (1.23) or (1.24). Notice that the model (1.26) is a special case of the *linear regression model with type II constrains* (1.27), where we set

$$Z = \begin{pmatrix} X_\Delta \\ Y_\Delta \end{pmatrix}, \gamma = \begin{pmatrix} \mu_\Delta \\ \nu_\Delta \end{pmatrix}, \theta = a_\Delta, A = A_0, B = B_0, c = c_0 \quad (1.34)$$

By using (1.30) we get the estimators of the calibration model parameters

$$\begin{pmatrix} \hat{\mu}_\Delta \\ \hat{\nu}_\Delta \end{pmatrix} = -\Sigma A_0'Q_{0,11}c_0 + (I - \Sigma A_0'Q_{0,11}A_0)\begin{pmatrix} X_\Delta \\ Y_\Delta \end{pmatrix}$$

$$\hat{a}_\Delta = -Q_{0,21}c_0 - Q_{0,21}A_0 \begin{pmatrix} X_\Delta \\ Y_\Delta \end{pmatrix}, \tag{1.35}$$

where $X_\Delta = X - \mu_0$, $Y_\Delta = Y - v_0$, and $Q_{0,11}$ and $Q_{0,12}$ are the blocks of the matrix Q_0 defined by

$$Q_0 = \begin{pmatrix} Q_{0,11} & Q_{0,12} \\ Q_{0,21} & Q_{0,22} \end{pmatrix} = \begin{pmatrix} A_0\Sigma A_0' & B_0 \\ B_0' & 0 \end{pmatrix}^{-1} \tag{1.36}$$

From (1.33), the covariance matrix of the (joint) locally best linear unbiased estimator is

$$Cov\left(\begin{pmatrix} \hat{\mu}_\Delta \\ \hat{v}_\Delta \\ \hat{a}_\Delta \end{pmatrix}\right) = \begin{pmatrix} \Sigma - \Sigma A_0' Q_{0,11} A_0\Sigma & -\Sigma A_0' Q_{0,12} \\ -Q_{0,21}A_0\Sigma & -Q_{0,22} \end{pmatrix} \tag{1.37}$$

From (1.22), (1.23) and/or (1.24) we get $A_0 = [\, D_0 \,\vdots\, -I\,]$ and moreover, from (1.8) we get that $\Sigma = Diag(\Sigma_X, \Sigma_Y)$ is a block diagonal matrix. Based on that, and from (1.35), we get the following expressions for the parameter estimators

$$\hat{\mu} = X - \Sigma_X D_0' Q_{0,11}(D_0 X_\Delta - Y)$$
$$\hat{v} = Y + \Sigma_Y Q_{0,11}(D_0 X_\Delta - Y)$$
$$\hat{a} = -(B_0' W_0^{-1} B_0)^{-1} B_0' W_0^{-1}(D_0 X_\Delta - Y), \tag{1.38}$$

where $W_0 = A_0\Sigma A_0' = D_0\Sigma_X D_0' + \Sigma_Y$.

The locally best linear unbiased estimators given in (1.38) can be further improved by imposing an iterative estimation procedure, such that the parameter estimate in the iteration step k is set as a prior value for the iteration step $k + 1$, i.e.

$$\mu_0^{(k+1)} = \hat{\mu}^{(k)}, v_0^{(k+1)} = \hat{v}^{(k)}, a_0^{(k+1)} = \hat{a}^{(k)} \tag{1.39}$$

until convergence is reached. Let μ_*, v_*, and a_* denote the pre-selected prior values of the parameters used in the final iteration step. Then the iterated BLUEs of the original calibration model parameters are

$$\hat{\mu} = X - \Sigma_X D_*' Q_{*,11}(D_* X_\Delta - Y)$$
$$\hat{v} = Y + \Sigma_Y Q_{*,11}(D_* X_\Delta - Y)$$
$$\hat{a} = -(B_*' W_*^{-1} B_*)^{-1} B_*' W_*^{-1}(D_* X_\Delta - Y) = L_X X_\Delta + K_Y Y, \tag{1.40}$$

where $X_\Delta = X - \mu_*$, $W_* = D_* \Sigma_X D'_* + \Sigma_Y$, $K_Y = (B'_* W_*^{-1} B_*)^{-1} B'_* W_*^{-1}$, and $L_X = -K_Y D_*$. The covariance matrix of the estimator is

$$Cov\left(\begin{pmatrix} \hat{\mu} \\ \hat{\nu} \\ \hat{a} \end{pmatrix}\right) = \begin{pmatrix} \Sigma - \Sigma A'_* Q_{*,11} A_* \Sigma & -\Sigma A'_* Q_{*,12} \\ -Q_{*,21} A_* \Sigma & -Q_{*,22} \end{pmatrix} \tag{1.41}$$

and in particular we get

$$Cov(\hat{a}) = \Sigma_{\hat{a}} = -Q_{*,22} = (B'_* W_*^{-1} B_*)^{-1} =$$

$$= L_X \Sigma_X L'_X + K_Y \Sigma_Y K'_Y. \tag{1.42}$$

1.4.2. Algorithm to Estimate the Calibration Model Parameters

Here we summarize the basic steps of the algorithm for evaluating the estimates of the calibration model parameters μ, ν, a and the estimate of its covariance matrix, based on the observed data $x = X^{(obs)}$ and $y = Y^{(obs)}$ (or the best estimates $\tilde{x} = (\tilde{x}_1, \ldots, \tilde{x}_I)'$ and $\tilde{y} = (\tilde{y}_1, \ldots, \tilde{y}_I)'$ specified by state-of-knowledge distributions in (1.2) and (1.4)).

(i) *Initialization:*

- Set the prior values of the model parameters μ_0, ν_0, a_0.

- Set the indicator of the iteration step, $k = 1$.

- Set the initial value of the criterion function, $crit = 1$.

- Set the limit for the number of iteration steps, e.g. $k_{max} = 100$.

- Set the preferred (small) tolerance value: $0 < \varepsilon \ll 1$.

- Set the starting values of the parameter estimates:

- $\hat{\mu}^{(0)} = \mu_0$. If not otherwise stated, one possible starting point is $\mu_0 = x$.

- $\hat{\nu}^{(0)} = \nu_0$. If not otherwise stated, one possible starting point is $\nu_0 = y$.

- $\hat{a}^{(0)} = a_0$. If not otherwise stated, one possible starting point is $a_0 = (M'_x M_x)^{-1} M'_x y$, where $M_x = [1, x, \ldots, x^p]$ (for the polynomial calibration function) or $\{M_x\}_{ij} = \sum_{k=0}^{q} A_{ij,k} \, x_i^k$, (for the generalized polynomial calibration function).

(ii) *Convergence:*

Check, if the following is true: $crit < \varepsilon$ or $k \geq k_{max}$.

If *true*, stop the algorithm:

- Save the estimated parameters $\hat{\mu}$, $\hat{v}$, $\hat{a}$, calculated by (1.40).

- Save the estimated covariance matrix $\Sigma_{\hat{a}}$, calculated by (1.42).

- Save the matrix $K_Y = (B'_* W_*^{-1} B_*)^{-1} B'_* W_*^{-1}$.

- Save the matrix D_* or $L_X = -K_Y D_* = -(B'_* W_*^{-1} B_*)^{-1} B'_* W_*^{-1} D_*$.

If *false*, continue and go to the next step (iii).

(iii) *Linearization:* For $k \geq 1$

- Set $\mu_0^{(k)} = \hat{\mu}^{(k-1)}$.

- Set $v_0^{(k)} = \hat{v}^{(k-1)}$.

- Set $a_0^{(k)} = \hat{a}^{(k-1)}$.

- Translate the observed data, $x_\Delta = x - \mu_0^{(k)}$, $y_\Delta = y - v_0^{(k)}$, and evaluate the required matrices A_0, B_0, and c_0 at $\mu_0 = \mu_0^{(k)}$, $v_0 = v_0^{(k)}$, and $a_0 = a_0^{(k)}$, as specified in (1.22), (1.23) or (1.24).

(iv) *Estimation:*

- Evaluate the estimates $\hat{\mu}$, $\hat{v}$ a $\hat{a}$ of the model parameters by using (1.38), for specified $\mu_0 = \mu_0^{(k)}$, $v_0 = v_0^{(k)}$, and $a_0 = a_0^{(k)}$.

(v) *Update:*

- Set $\hat{\mu}^{(k+1)} = \hat{\mu}$.

- Set $\hat{v}^{(k+1)} = \hat{v}$.

- Set $\hat{a}^{(k+1)} = \hat{a}$.

- Evaluate the criterion

$$crit = \frac{1}{I}\left(\left|\hat{\mu}^{(k+1)} - \hat{\mu}^{(k)}\right|^2 + \left|\hat{v}^{(k+1)} - \hat{v}^{(k)}\right|^2 + \left|\hat{a}^{(k+1)} - \hat{a}^{(k)}\right|^2 \right), \quad (1.43)$$

where by $|v|^2$ we denote the Euclidean norm of a vector v.

- Update the iteration step, set $k = k + 1$, and go to (ii) to check the convergence.

1.5. State-of-Knowledge Distribution about the Calibration Parameters

Based on (1.40), the iterative BLUE (best linear unbiased estimator) of the vector of the calibration function parameters $a = (a_0, \ldots, a_p)'$, is given by

$$\hat{a} = -(B_*'W_*^{-1}B_*)^{-1}B_*'W_*^{-1}(D_*(X - \mu_*) - Y) =$$

$$= L_X(X - \mu_*) + K_Y Y, \tag{1.44}$$

where the random vectors X and Y are specified by (1.1) and (1.3), and represent the used measurement process.

Given the state-of-knowledge distributions $\tilde{\mu}_i \sim \tilde{x}_i - \tilde{X}_i$ and $\tilde{v}_i \sim \tilde{y}_i - \tilde{Y}_i$, as specified in (1.2) and (1.4) based on using the direct measurements of μ_i and v_i for $i = 1, \ldots, I$, the *best estimate* of the calibration parameter $a = (a_0, \ldots, a_p)'$ is given by

$$\bar{a} = L_X(\tilde{x} - \mu_*) + K_Y \tilde{y}, \tag{1.45}$$

where $\tilde{x} = x - \delta_x$ and $\tilde{y} = y - \delta_y$ are nonstochastic values considered as the *best estimates* of μ and v, based on the direct measurements and the expert knowledge about systematic biases.

Then, the state-of-knowledge distribution about the parameters $a = (a_0, \ldots, a_p)'$ is the distribution of the random variable $\tilde{a}$, specified by

$$\tilde{a} \sim \bar{a} - \left(L_X \tilde{X} + K_Y \tilde{Y}\right), \tag{1.46}$$

where $\tilde{X} \sim Z_X + \Delta_X + \nabla_X 1$ and $\tilde{Y} \sim Z_Y + \Delta_Y + \nabla_Y 1$ and all random vectors are mutually independent with known distributions, specified by (1.1), (1.3) and (1.6).

Note that $\bar{a}$ is the non-stochastic part of $\tilde{a}$, representing our *best estimate* of the calibration parameter a, and $L_X\tilde{X} + K_Y\tilde{Y}$ is the stochastic part of $\tilde{a}$, with zero-mean probability distribution, characterising our knowledge about the associated uncertainty.

In particular, the marginal state-of-knowledge distribution about the parameter a_k, for a specific $k = 0, \dots, p$, is

$$\tilde{a}_k = \bar{a}_k - \left(L_{X,k}\tilde{X} + K_{Y,k}\tilde{Y}\right) = \bar{a}_k - L_{X,k}(Z_X + \Delta_X + \nabla_X 1) -$$
$$-K_{Y,k}\left(Z_Y + \Delta_Y + \nabla_Y 1\right) = \bar{a}_k - L_{X,k}Z_X - L_{X,k}\Delta_X -$$
$$-\left(L_{X,k}1\right)\nabla_X - K_{Y,k}Z_Y - K_{Y,k}\Delta_Y - \left(K_{Y,k}\,1\right)\nabla_Y, \qquad (1.47)$$

where $\bar{a}_k$ is the *best estimate* of a_k, the $(k+1)^{\text{th}}$ component of the vector $\bar{a}$, and $L_{X,k}$, $K_{Y,k}$, are the $(k+1)$th rows of the matrices L_X and K_Y, respectively, for $k = 0, \dots, p$.

Moreover, the state-of-knowledge distribution about the linear combination of the calibration function parameters, say $w'a = \sum_{k=0}^{p} w_{k+1}a_k$, is

$$\tilde{g} \sim w'\bar{a} - \left(w'L_X\tilde{X} + w'K_Y\tilde{Y}\right) = w'\bar{a} - w'L_X(Z_X + \Delta_X + \nabla_X 1) -$$
$$-w'K_Y\left(Z_Y + \Delta_Y + \nabla_Y 1\right) = w'\bar{a} - (w'L_X)Z_X - (w'L_X)\Delta_X -$$
$$-(w'L_X 1)\nabla_X - (w'K_Y)Z_Y - (w'K_Y)\Delta_Y - (w'K_Y 1)\nabla_Y = w'\bar{a} -$$
$$-w'_X Z_X - w'_X\Delta_X - (w'_X 1)\nabla_X - w'_Y Z_Y - w'_Y\Delta_Y - (w'_Y 1)\nabla_Y, \quad (1.48)$$

where $w = \left(w_1, \dots, w_{p+1}\right)'$ are the coefficients of the considered linear combination, and $w_X = (w'L_X)' = \left(w_{X,1}, \dots, w_{X,I}\right)'$ and $w_Y = (w'K_Y)' = \left(w_{Y,1}, \dots, w_{Y,I}\right)'$ are the associated coefficients standing with the random variables $\tilde{X}$ and $\tilde{Y}$.

Frequently, $w = (1, x, \dots, x^p)'$ for some pre-selected scalar value x, specified from the considered range of values of the measuring instrument X.

The explicit expressions for the state-of-knowledge distributions about the calibration parameters $a = (a_0, \dots, a_p)'$ or their linear combinations, as specified in (1.46), (1.47) or (1.48), are difficult to derive.

However, such distributions can be reasonably well estimated by numerical computer intensive methods, e.g. by using the approach based on the *Monte Carlo Methods (MCM)*, as proposed in the *GUM Supplements 1 and 2*, see [3] and [4], or by using the *Characteristic Functions Approach (CFA)*, as proposed in [6]. For more details about principles of uncertainty analysis, based on using the state-of-knowledge distribution, and methods used for evaluation of such distributions in the frame of the standard univariate (linear) measurement models see the Appendix.

In general, the MCM approach is based on a much simpler concept and the required algorithms are relatively easy to implement with the standard computer programming environments for technical computing as MATLAB or R. However, when the state-of-knowledge distribution is required to be evaluated with high precision for a large number of different functions (linear combinations of the calibration parameters), the required computational burden can be substantial.

On the other hand, algorithmic implementation of the CFA requires more advanced tools, which are available but still under development, for more details see e.g. [7]. As argued in [6], CFA allows highly precise and efficient computation of the required tail probabilities, used for uncertainty analysis in calibration and computing the required coverage intervals.

1.6. Measuring with the Calibrated Device

In probability and statistics, the Bonferroni method is a simple conservative method that allows making probability statements and/or constructing probability intervals (confidence, prediction and/or coverage intervals) while assuring that the pre-specified overall probability level is maintained. Formally, the general Bonferroni inequality is presented by:

$$Prob(\cap_{i=1}^{n} E_i) \geq 1 - \sum_{i=1}^{n} Prob(E_i^c), \qquad (1.49)$$

where E_i and its complement E_i^c are any random events.

Let us consider the generalized polynomial calibration function, $v = f(\mu, a)$, specified by its parameters $a = (a_0, ..., a_p)'$. For every fixed μ we get

$$v_\mu = f(\mu, a) = w'_\mu a = \sum_{k=0}^{p} w_{\mu,k+1} a_k, \qquad (1.50)$$

where $w_\mu = (w_{\mu,1}, \ldots, w_{\mu,p+1})'$ is a vector of known coefficients of the considered linear combination. So, we can specify the state-of-knowledge distribution about $v_\mu = w'_\mu a$ for any given μ. Hence, for pre-specified level $\alpha \in (0,1)$, we can construct the $(1 - \alpha/2)$-coverage interval for any v_μ, say $(L_{\mu,1-\alpha/2}, U_{\mu,1-\alpha/2})$, such that

$$Prob\left(v_\mu \in \left(L_{\mu,1-\frac{\alpha}{2}}, U_{\mu,1-\frac{\alpha}{2}}\right)\right) = 1 - \frac{\alpha}{2} \qquad (1.51)$$

Let $S_{1-\alpha/2}$ denotes the region of values (μ, v) about the theoretical calibration function, defined by

$$S_{1-\frac{\alpha}{2}} = \left\{(\mu, v): v = w'_\mu a \ \& v \in \left(L_{\mu,1-\frac{\alpha}{2}}, U_{\mu,1-\frac{\alpha}{2}}\right) \text{ for all } \mu\right\}, \ (1.52)$$

where we further assume that for the considered range of μ values the *calibration function is strictly monotone*.

On the other hand, for any fixed value v (at the y-axis) we can construct an interval of μ values, say $\mu_v \in (L_{v,1-\alpha/2}, U_{v,1-\alpha/2})$, such that is given as the intersection of the region $S_{1-\alpha/2}$ and the horizontal line at the level v (i.e. line crossing the point $(0, v)$). From the given construction it is clear that for any v the following probability statetment is true

$$Prob\left(\mu_v \in \left(L_{v,1-\frac{\alpha}{2}}, U_{v,1-\frac{\alpha}{2}}\right)\right) = 1 - \frac{\alpha}{2} \qquad (1.53)$$

Now, let us assume that we have a new measurement of v, which is fully specified by its state-of-knowledge distribution. Hence, we can contruct the $(1 - \alpha/2)$-coverage interval for the values of v, say $(v_{L,1-\alpha/2}, v_{U,1-\alpha/2})$, such that

$$Prob\left(v \in \left(v_{L,1-\frac{\alpha}{2}}, v_{U,1-\frac{\alpha}{2}}\right)\right) = 1 - \frac{\alpha}{2} \qquad (1.54)$$

Finally, let R denotes a region of values (μ, v), constructed by intersecting the strip region $S_{1-\alpha/2}$ and the (μ, v) values from the belt restricted by $v \in (v_{L,1-\alpha/2}, v_{U,1-\alpha/2})$,

$$R = \left\{(\mu, v): (\mu, v) \in S_{1-\frac{\alpha}{2}} \ \& \ v \in \left(v_{L,1-\frac{\alpha}{2}}, v_{U,1-\frac{\alpha}{2}}\right)\right\} \quad (1.55)$$

By using the Bonferroni inequality (1.49) we get

$$Prob\big((\mu, v) \in R\big) \geq 1 - \alpha \quad (1.56)$$

and hence, the interval for the values of μ, specified by limits of the region R, say (μ_L, μ_U), where

$$\mu_L = \min_{(\mu,v) \in R} \mu, \mu_U = \max_{(\mu,v) \in R} \mu \quad (1.57)$$

is the Bonferroni-type $(1 - \alpha)$-coverage interval for μ such that

$$Prob\big(\mu \in (\mu_L, \mu_U)\big) \geq 1 - \alpha \quad (1.58)$$

For more details see Fig. 1.1, which illustrates the principle of measuring with the calibrated device and construction of the Bonferroni-type coverage interval for μ based on the information (state-of-knowledge distribution and/or the coverage interval) about v from the measurements.

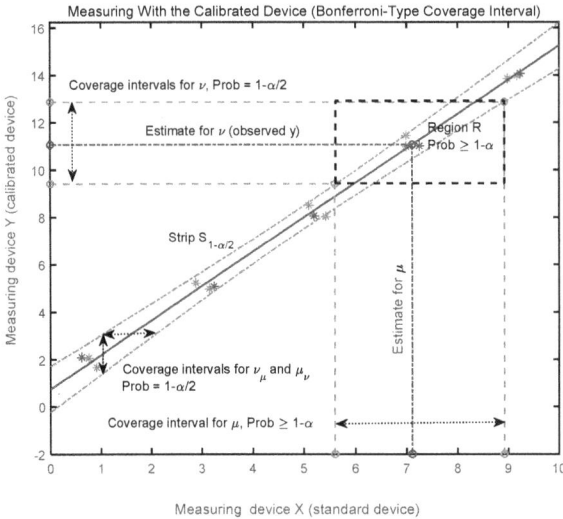

Fig. 1.1. Illustration of the principle of measuring with the calibrated device and the construction of the Bonferroni-type coverage interval for μ based on the information (state-of-knowledge distribution and/or the coverage interval) about v from the measurements.

1.7. Example: Calibration of the Industrial Platinum Resistance Thermometer

Application of the industrial platinum resistance temperature sensors for temperature measurements is based on a unique relationship between temperature and resistance. There is exactly one resistance value for each temperature. This clear relationship can be described by mathematical formulae. The international standard IEC 60751:2008 [9] specifies the requirements and temperature/resistance relationship for industrial platinum resistance temperature sensors and industrial platinum resistance thermometers whose electrical resistance is a defined function of temperature.

According to [9] the formulae apply within the operating temperature limits and do not depend on the design of the measuring resistor. Based on equations suggested by Callendar and Van Dusen, see [8], the following applies in the temperature range from -200 to $0\,°C$:

$$R_t = R_0(1 + At + Bt^2 + C(t - 100)t^3) \qquad (1.59)$$

The following applies in the temperature range from 0 to $+600\,°C$:

$$R_t = R_0(1 + At + Bt^2), \qquad (1.60)$$

where t denotes the temperature in degrees of celsius ($°C$), R_t denotes the resistance in ohms (Ω) at the measured temperature, R_0 denotes the resistance in ohms at $t = 0\,°C$ (e.g. $100\,\Omega$). According to [9], the following constants apply for the purpose of the calculation:

$$R_0 = 100,\ A = 3.9083 \times 10^{-3},\ B = -5.7750 \times 10^{-7},$$

$$C = -4.1830 \times 10^{-12} \qquad (1.61)$$

It is important to note that these equations and the constants in (1.61) are listed as the basis for the temperature/resistance tables for *idealized platinum resistance thermometers* and are not intended to be used for the calibration of an individual thermometer, which require the experimentally determined parameters to be found.

For illustration purposes, here we consider comparative calibration of the industrial platinum resistance temperature sensor by using the standard platinum resistance thermometer, based on data from the calibration experiment realized at the Slovak Institute of Metrology (SMU) in Bratislava, Slovakia.

In fact, it is a comparative calibration problem of two measurement instruments X and Y. The more precise instrument denoted as X represents here the *Standard Platinum Resistance Thermometer* (SPRT) and the less precise instrument, denoted as Y, represents the *Industrial Platinum Resistance Thermometer* (IPRT). The calibration experiment was realized by $n = 100$ replicated measurements performed on a set of $I = 11$ suitably chosen quantities – a pre-selected temperature states (with nominal temperatures -20, -10, 0, 10, 20, 30, 40, 50, 60, 70, and 80 °C). The SMU uncertainty budget for temperature measured by SPRT is given in Table 1.1, the SMU uncertainty budget for resistance measured by the calibrated IPRT is given in Table 1.2.

For SPRT, the direct measurement process for measuring the ith temperature $(i = 1, \ldots, I)$ is assumed to be modelled by the following measurement equation

$$\mu_i = t_i - t_{SPRT,i} - \Delta_{SPRT,i} - \nabla_{SPRT} =$$

$$= t_i - \left(t_{SPRT,i} + \delta t_{hom,i} + \delta t_{stab,i}\right) -$$

$$- (\delta R_{S1} + \delta R_{S2} + \delta R_{S3} + \delta R_{DSPRT} + \delta R_{DB}), \quad (1.62)$$

where

- t_i is the best estimate of the temperature measured by SPRT, $i = 1, \ldots, I$.

- $t_{SPRT,i}$ is the zero-mean random variable representing knowledge about the measurement errors, with the type A standard deviations $u_{t_{SPRT,i}}$, as presented in Table 1.1 for all $i = 1, \ldots, I$. Here we assume that $t_{SPRT,i} = u_{t_{SPRT,i}} Z_i$, where $Z_i \sim N(0,1)$ are independent Gaussian random variables for $i = 1, \ldots, I$.

- $\delta t_{hom,i}$ is the independent random variable representing knowledge about the necessary correction on homogeneity of bath, with known zero-mean rectangular distribution with given standard deviation $u_{\delta t_{hom,i}}$, given in Table 1.1 from type B evaluations, for $i = 1, \ldots, I$, i.e. $\delta t_{hom,i} \sim R(-\sqrt{3} u_{\delta t_{hom,i}}, \sqrt{3} u_{\delta t_{hom,i}})$,

Table 1.1. Uncertainty budget for temperature measured by SPRT in degrees of Celsius (°C). The presented uncertainties are in milikelvins (mK). The uncertainties depicted in the grey sector are common for all measurements (i.e. they are related to the state-of-knowledge distributions of the random variables, common for all measurements).

est estimate	uncertainty from SPRT measurements	SMU uncertainty budget for temperature measured by the standard SPRT (°C)							combined uncertainty
		uncertainty from temperature gradient in water bath (homogeneity)	uncertainty from temperature stability within the water bath	uncertainty from calibration certificate of the SPRT	uncertainty from resolution of a displaying device of the SPRT	uncertainty from calibration certificate of a displaying device of the SPRT	uncertainty from drift of the SPRT	uncertainty from drift of a displaying device of the SPRT	
t	t_{SPRT}	δt_{hom}	δt_{stab}	δR_{S1}	δR_{S2}	δR_{S3}	δR_{DSPRT}	δR_{DB}	
	type A	type B	type B	type B	type B	type B	type B	type B	
	u_{tSPRT}	$u_{\delta hom}$	$u_{\delta stab}$	$u_{\delta RS1}$	$u_{\delta RS2}$	$u_{\delta RS3}$	$u_{\delta RDSPRT}$	$u_{\delta RDB}$	
	Gaussian	Rectangular	Rectangular	Gaussian	Rectangular	Rectangular	Gaussian	Rectangular	
Value	mK	mK	mK	mK	mK	mK	mK	mK	mK
°C									
-19.68434	1.20	9.81	8.66	1.50	0.29	2.50	1.00	1.00	**13.54**
-9.81417	1.50	6.35	9.81	1.50	0.29	2.50	1.00	1.00	**12.23**
0.09888	1.50	5.77	9.24	1.50	0.29	2.50	1.00	1.00	**11.47**
10.01491	1.40	5.77	9.81	1.50	0.29	2.50	1.00	1.00	**11.93**
19.96336	2.00	6.35	8.66	1.50	0.29	2.50	1.00	1.00	**11.40**
29.81308	1.00	8.08	8.08	1.50	0.29	2.50	1.00	1.00	**11.93**
39.84057	1.10	10.39	8.66	1.50	0.29	2.50	1.00	1.00	**13.96**
49.80648	0.80	12.70	9.24	1.50	0.29	2.50	1.00	1.00	**16.06**
59.85598	1.40	13.86	8.66	1.50	0.29	2.50	1.00	1.00	**14.86**
69.99034	2.10	13.86	9.24	1.50	0.29	2.50	1.00	1.00	**17.10**
79.89018	1.90	15.01	8.66	1.50	0.29	2.50	1.00	1.00	**17.73**

Table 1.2. Uncertainty budget for electrical resistance measured by the calibrated PRT in ohms (Ω). The presented uncertainties are in milliohms (mΩ). The uncertainties depicted in the grey sector are common for all measurements (i.e. they are related to the state-of-knowledge distributions of the random variables, common for all measurements).

SMU uncertainty budget for resistance measured by the calibrated IPRT (Ω)						
best estimate	uncertainty from IPRT measurements	uncertainty from temperature determination of the bath measured by	uncertainty from resolution of the IPRT multimeter	uncertainty from calibration certificate of the IPRT multimeter	uncertainty from drift of the IPRT multimeter	combined uncertainty
R	RI_{PRT}	δt_X	δR_R	δR_{CM}	δR_{DM}	
	type A	type B	type B	type B	type B	
	u_{RIPRT}	$u_{\delta tx}$	$u_{\delta RR}$	$u_{\delta RCM}$	$u_{\delta RDM}$	
Value	Gaussian	Gaussian	Rectangular	Gaussian	Rectangular	
Ω	mΩ	mΩ	mΩ	mΩ	mΩ	mΩ
92.14543	0.60	5.21	0.29	1.00	0.33	5.36
96.09286	0.80	4.71	0.29	1.00	0.33	4.90
100.04613	0.80	4.41	0.29	1.00	0.33	4.62
103.98846	0.60	4.59	0.29	1.00	0.33	4.76
107.93124	0.70	4.39	0.29	1.00	0.33	4.58
111.82728	0.50	4.59	0.29	1.00	0.33	4.75
115.77530	0.60	5.37	0.29	1.00	0.33	5.52
119.69136	0.40	6.18	0.29	1.00	0.33	6.29
123.62541	0.70	6.44	0.29	1.00	0.33	5.87
127.58131	1.00	6.58	0.29	1.00	0.33	6.75
131.43388	1.00	6.83	0.29	1.00	0.33	6.99

- $\delta t_{stab,i}$ is the independent random variable representing knowledge about the necessary correction on temperature gradient in water bath, i.e. stability of bath, with known zero-mean rectangular distribution with given standard deviation $u_{\delta t_{stab,i}}$, given in Table 1.1 from type B evaluations, i.e. $\delta t_{stab,i} \sim R(-\sqrt{3}u_{\delta t_{stab,i}}, \sqrt{3}u_{\delta t_{stab,i}})$, for $i = 1, \dots, I$.

- δR_{S1} is the random variable representing knowledge about the uncertainty from calibration certificate of the SPRT, with known zero-mean rectangular distribution with type B standard deviation $u_{\delta R_{S1}}$, given in Table 1.1. Here we assume that $\delta R_{S1} = u_{\delta R_{S1}} Z$, where $Z \sim N(0,1)$ is a Gaussian random variable common for all $i = 1, \dots, I$.

- δR_{S2} is the random variable representing knowledge about the uncertainty from resolution of a displaying device of the SPRT (resolution of the bridge) with known zero-mean rectangular distribution with type B standard deviation $u_{\delta R_{S2}}$, given in Table 1.1, i.e. $\delta R_{S2} \sim R(-\sqrt{3}u_{\delta R_{S2}}, \sqrt{3}u_{\delta R_{S2}})$ is a common random variable for all $i = 1, \dots, I$.

- δR_{S3} is the random variable representing knowledge about uncertainty from calibration certificate of a displaying device of the SPRT (calibration uncertainty of the bridge) with known zero-mean rectangular distribution with type B standard deviation $u_{\delta R_{S3}}$, given in Table 1.1, i.e. $\delta R_{S3} \sim R(-\sqrt{3}u_{\delta R_{S3}}, \sqrt{3}u_{\delta R_{S3}})$, is a common random variable for all $i = 1, \dots, I$.

- δR_{DSPRT} is the random variable representing knowledge about uncertainty from drift of the SPRT (SPRT drift), with known zero-mean rectangular distribution with type B standard deviation $u_{\delta R_{DSPRT}}$, given in Table 1.1. Here we assume that $\delta R_{DSPRT} = u_{\delta R_{DSPRT}} Z$, where $Z \sim N(0,1)$ is a Gaussian random variable common for all $i = 1, \dots, I$.

- δR_{DB} is the random variable representing knowledge about uncertainty from drift of a displaying device of the SPRT (drift of the bridge) with known zero-mean rectangular distribution with type B standard deviation $u_{\delta R_{DB}}$, given in Table 1.1, i.e. $\delta R_{DB} \sim R(-\sqrt{3}u_{\delta R_{DB}}, \sqrt{3}u_{\delta R_{DB}})$, is a common random variable for all $i = 1, \dots, I$.

For the calibrated IPRT, the direct measurement process for measuring the ith resistance ($i = 1, \dots, I$) is assumed to be modelled by the following measurement equation

$$v_i = R_i - R_{IPRT,i} - \Delta_{IPRT,i} - \nabla_{IPRT} =$$

$$= R_i - \left(R_{IPRT,i} + \delta t_{X,i}\right) - \left(\delta R_R + \delta R_{CM} + \delta R_{DM}\right), \quad (1.63)$$

where

- R_i is the best estimate of the resistance measured by IPRT, $i = 1, \dots, I$.

- $R_{IPRT,i}$ is the zero-mean random variable representing knowledge about the measurement errors, with the type A standard deviations $u_{R_{IPRT,i}}$, as presented in Table 1.2 for all $i = 1, \dots, I$. Here we assume that $R_{IPRT,i} = u_{R_{IPRT,i}} Z_i$, where $Z_i \sim N(0,1)$ are independent Gaussian random variables for $i = 1, \dots, I$.

- $\delta t_{X,i}$ is the zero-mean random variable representing knowledge about the uncertainty from temperature determination of the bath (temperature of the bath), with the type A standard deviation $u_{\delta t_{X,i}}$, as presented in Table 1.2 for all $i = 1, \dots, I$. Here we assume that $\delta t_{X,i} = u_{\delta t_{X,i}} Z_i$, where $Z_i \sim N(0,1)$ are independent Gaussian random variables for $i = 1, \dots, I$.

- δR_R is the random variable representing knowledge about the uncertainty from resolution of the IPRT multimeter (resolution of the multimeter), with known zero-mean rectangular distribution with type B standard deviation $u_{\delta R_R}$, given in Table 1.2. Here we assume that $\delta R_{S1} = u_{\delta R_{S1}} Z$, where $Z \sim N(0,1)$ is a Gaussian random variable common for all $i = 1, \dots, I$.

- δR_{CM} is the random variable representing knowledge about uncertainty from calibration certificate of the IPRT multimeter (calibration uncertainty of the multimeter) with known zero-mean rectangular distribution with type B standard deviation $u_{\delta R_{CM}}$, given in Table 1.2, i.e. $\delta R_{CM} \sim R(-\sqrt{3}u_{\delta R_{CM}}, \sqrt{3}u_{\delta R_{MC}})$, is a common random variable for all $i = 1, \dots, I$.

- δR_{DM} is the random variable representing knowledge about uncertainty from drift of the IPRT multimeter (drift of the multimeter) with known zero-mean rectangular distribution with type B standard deviation $u_{\delta R_{DM}}$, given in Table 1.2, i.e. $\delta R_{DM} \sim R(-\sqrt{3}u_{\delta R_{DM}}, \sqrt{3}u_{\delta R_{DM}})$, is a common random variable for all $i = 1, \dots, I$.

Here we consider comparative calibration model with the generalized polynomial calibration function, specified by

$$v_i = a_0 + a_1\mu_i + a_2\mu_i^2 + a_3(\mu_i - 100)\mu_i^3 \text{ for } i = 1,2,$$

$$v_i = a_0 + a_1\mu_i + a_2\mu_i^2 \text{ for } i = 3, \dots, 11, \qquad (1.64)$$

where v_i and μ_i represet the IPRT resistances and the SPRT temperatures, respectively, at the pre-specified measurement points, $i = 1, \dots, 11$. The parameters $a = (a_0, \dots, a_3)'$ of the generalized polynomial calibration function are uniquely related to the original parametrization (1.59) and (1.60), suggested in [9]. In particular

$$(R_0, A, B, C) = \left(a_0, \frac{a_1}{a_0}, \frac{a_2}{a_0}, \frac{a_3}{a_0}\right) \tag{1.65}$$

For the data specified by the uncertainty budgets in Table 1.1 and Table 1.2, by using (1.40) and (1.41) we get the following best estimates of the parameters $a = (a_0, \dots, a_3)'$ and their uncertainties

$$\hat{a}_0 = 100.0105 \; (4.0057 \times 10^{-3})$$

$$\hat{a}_1 = 3.9825 \times 10^{-1} \; (2.4541 \times 10^{-4})$$

$$\hat{a}_2 = -6.0804 \times 10^{-5} (3.4570 \times 10^{-6})$$

$$\hat{a}_3 = 9.1845 \times 10^{-9} \; (7.6764 \times 10^{-8}) \tag{1.66}$$

The estimated covariance matrix (1.42) is

$$\Sigma_{\hat{a}} = \begin{pmatrix} 1.61 \times 10^{-5} & -5.18 \times 10^{-7} & 4.54 \times 10^{-9} & -1.74 \times 10^{-10} \\ -5.18 \times 10^{-7} & 6.02 \times 10^{-8} & -7.92 \times 10^{-10} & 1.38 \times 10^{-11} \\ 4.54 \times 10^{-9} & -7.92 \times 10^{-10} & 1.20 \times 10^{-11} & -1.69 \times 10^{-13} \\ -1.74 \times 10^{-10} & 1.38 \times 10^{-11} & -1.69 \times 10^{-13} & 5.89 \times 10^{-15} \end{pmatrix} \tag{1.67}$$

The marginal state-of-knowledge distributions of the parameters, calculated by using (1.47) and the characteristic functions approach, are depicted in Fig. 1.2.

For comparison with the standard values (1.61) we present the calibration estimates

$$R_0 = 100.0105 \, , A = 3.9821 \times 10^{-3},$$

$$B = -6.0798 \times 10^{-7}, C = 9.1835 \times 10^{-11} \tag{1.68}$$

Notice that based on our calibration experiment and the calculated results, the parameter a_3, and hence also the reparametrized parameter C, is statistically nonsignificant.

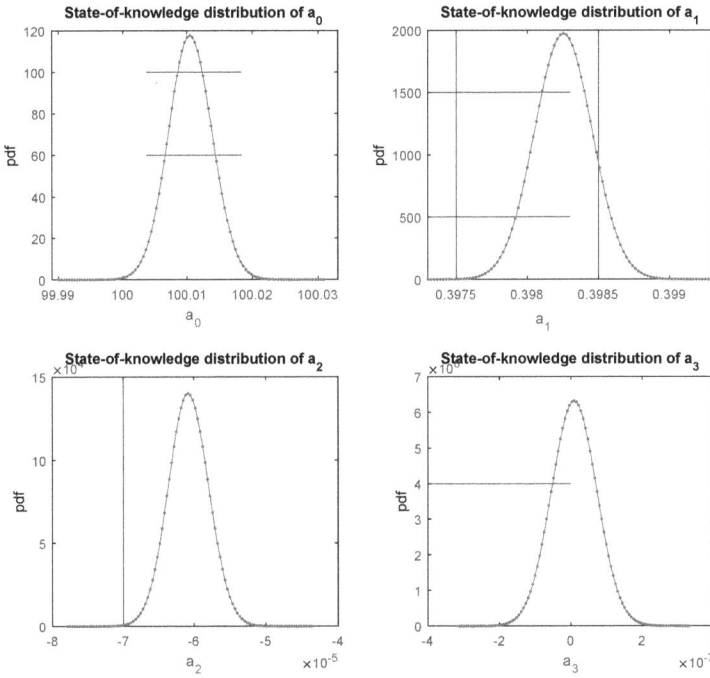

Fig. 1.2. The evaluated marginal state-of-knowledge distributions (the probability density functions) of the generalized polynomial calibration function parameters $a = (a_0, \ldots, a_3)'$.

Finally, in order to illustrate measuring with the calibrated device (IPRT) let us assume, that for $\alpha = 0.05$ the $(1 - \frac{\alpha}{2})$-coverage interval for v, derived from state-of-knowledge distribution of the new measurements, is specified by the following limits

$$\left(v_{L,1-\frac{\alpha}{2}}, v_{U,1-\frac{\alpha}{2}}\right) = (103.8998 \ \Omega, 104.0850 \ \Omega) \qquad (1.69)$$

Hence, by using the Bonferroni-type construction of the associated coverage intervals for μ, as presented in Section 1.6, from (1.57) we get the conservative estimate of the $(1 - \alpha)$-coverage interval for μ

$$(\mu_L, \mu_U) = (9.7645 \ °C, 10.2632 \ °C) \qquad (1.70)$$

such that $Prob\big(\mu \in (9.7645 \ °C, 10.2632 \ °C)\big) \geq 1 - \alpha$

43

1.8. Conclusions

In this chapter we have derived methods and algorithms, based on using the classical regression approach used for the errors-in-variable model and applied it for the comparative calibration with the generalized polynomial calibration function. The present approach is based on strict metrology considerations. The inputs quantities are fully characterized by their state-of-knowledge distributions, derived from the associated uncertainty budgets of the direct measurements from the calibration experiment, which allows to combine type A and type B methods of evaluation of the measurements, as well as to incorporate correlated measurements.

The suggested methods are computationally intensive and can be based on using the Monte Carlo methods (MCM), as proposed in the GUM Supplements 1 and 2, see [3] and [4], or by using the characteristic functions approach (CFA), as proposed in [6].

One drawback of the classical regression approach to calibration (i.e. method based on modelling y-values as a function of x-values in the standard x–y space) is the inherent complication in deriving the associated state-of-knowlege distribution for μ based on the new (independent) measurements with the calibrated device. Because of that, we have suggested approximate and rather conservative Bonferroni-type method for construction of the associated coverage intervals for μ.

A possible remedy for the problem is to consider the inverse regression approach (i.e. method based on modelling x-values as a function of y-values in the standard x–y space). In a seminal paper [10], Krutchkoff observed that application of the inverse regression approach for linear calibration has uniformly smaller mean squared error than the classical approach. Although the classical and inverse polynomial regression functions are not equivalent modelling functions, if there is no other serious reason (as e.g. the authoritative specification by the international standard documents) we prefer to use the inverse regression approach for polynomial calibration, i.e. modelling x-values as a function of y-values in the standard x–y space. In particular, as an alternative to (1.12), we suggest to consider the following polynomial calibration function of order p,

$$\mu = a_0 1 + a_1 v + \cdots + a_p v^p = [1, \, v, \cdots, v^p] \begin{pmatrix} a_0 \\ a_1 \\ \vdots \\ a_p \end{pmatrix} = Na \quad (1.71)$$

Such change is easily applicable for the errors-in-variable models by simple renaming of the input variables ($x \leftrightarrow y$, $\mu \leftrightarrow v$) in the already presented formulas and algorithms. The important benefit of this change is that it allows full specification and evaluation of the exact state-of-knowledge distribution for μ, based on the state-of-knowledge distribution for v, from the new observation with the calibrated device.

Another problem which remains open is related to numerical issues associated with fitting the standard polynomial functions. As we have illustrated by the example on calibration of the industrial platinum resistance thermometer, the numerical values of the appropriate polynomial coefficient are very different (by several orders of magnitude). As suggested by Cox and Harris in [11], in such situations it is better to consider polynomial modelling based on using the orthogonal Chebyshcv polynomials or other type of standardized polynomials. Application of the approach based on Chebyshev polynomials for the presented calibration methods is possible, but still open for future research.

Acknowledgements

The authors would like to thank Prof. Stanislav Ďuriš and Prof. Rudolf Palenčár of the Slovak Technical University (SjF STU) and Dr. Milan Ioan Maniur and Dr. Zuzana Ďurišová of the Slovak Institute of Metrology (SMÚ) for providing the experimental calibration data of the Industrial Platinum Resistance Thermometer, and their helpful advice on various technical issues examined in this chapter. The work was supported by the Slovak Research and Development Agency, project APVV-15-0295, and by the Scientific Grant Agency VEGA of the Ministry of Education of the Slovak Republic and the Slovak Academy of Sciences, by the projects VEGA 2/0054/18 and VEGA 2/0011/16.

References

[1]. JCGM200:2012, The International Vocabulary of Metrology – Basic and General Concepts And Associated Terms (VIM), 3rd Ed., *JCGM – Joint Committee for Guides in Metrology*, https://goo.gl/ZaUMCY

[2]. JCGM100:2008, Evaluation of Measurement Data – Guide to the Expression of Uncertainty in Measurement, *JCGM – Joint Committee for Guides in Metrology*, https://goo.gl/mBMa5d

[3]. JCGM101:2008, Evaluation of Measurement Data – Supplement 1 to the Guide to the Expression of Uncertainty in Measurement – Propagation of Distributions Using a Monte Carlo Method, *JCGM – Joint Committee for Guides in Metrology*, https://goo.gl/8L2tE3

[4]. JCGM102:2011, Evaluation of Measurement Data – Supplement 2 to the Guide to the Expression of Uncertainty in Measurement – Extension To Any Number of Output Quantities, *JCGM – Joint Committee for Guides in Metrology*, https://goo.gl/hmhRS1

[5]. L. Kubáček, Foundations of Estimation Theory, *Elsevier*, 1988.

[6]. V. Witkovský, Numerical inversion of a characteristic function: An alternative tool to form the probability distribution of output quantity in linear measurement models, *Acta IMEKO*, Vol. 5, Issue 3, 2016, pp. 32-44.

[7]. V. Witkovský, CharFunTool: The characteristic functions toolbox, in *Proceedings of the 10th International Conference of the ERCIM Workgroup on Computational and Methodological Statistics (CMStatistics'17)*, Senate House, University of London, UK, 16-18 December 2017, 126.

[8]. M. S. Van Dusen, Platinum-resistance thermometry at low temperatures, *Journal of the American Chemical Society*, Vol. 47, Issue 2, 1925, pp. 326-332.

[9]. IEC 60751:2008. Industrial Platinum Resistance Thermometers and Platinum Temperature Sensors, *International Electrotechnical Commission*, https://goo.gl/WUibsk

[10]. R. G. Krutchkoff, Classical and inverse regression methods of calibration, *Technometrics*, Vol. 9, Issue 3, 1967, pp. 425-439.

[11]. M. Cox, P. Harris, Polynomial calibration functions revisited: Numerical and statistical issues, *Advanced Mathematical and Computational Tools in Metrology and Testing X*, 2015, pp. 9-16.

[12]. V. Witkovský, G. Wimmer, Z. Ďurišová, S. Ďuriš, R. Palenčár, Brief overview of methods for measurement uncertainty analysis: GUM uncertainty framework, Monte Carlo method, characteristic function approach, in *Proceedings of the IEEE 11th International Conference on Measurement (Measurement'17)*, 2017, pp. 35-38.

Appendix 1.A

The Appendix presents basic information about principles of uncertainty analysis, based on using the state-of-knowledge distribution, and methods used for evaluation of such distributions in the frame of the standard univariate (linear) measurement models. The following text is adapted from [6] and [8].

1.A.1. Measurement Uncertainty and the State-of-Knowledge Distributions

In metrology, *measurement uncertainty* is a non-negative parameter characterizing the dispersion of the values attributed to a measured quantity.

The basic working tool in measurement uncertainty analysis, as advocated in the *Guide to the expression of uncertainty in measurement* (GUM) [2], and in particular by its *Supplements* [3] and [4], is the *state-of-knowledge distribution* about the quantity derived based on the currently available information. Hence, the *measurement uncertainty* is taken as the *standard deviation of a state-of-knowledge probability distribution* over the possible values that could be attributed to a measured quantity.

For practical purposes, often is required an interval (typically as small as possible) containing the attributed quantity values with a specified (high) probability. Such an interval, a *coverage interval*, can be deduced from the *state-of-knowledge* probability distribution for the measured quantity. The specified probability is known as the *coverage probability*.

In general, for a given coverage probability there is more than one coverage interval. However, properly derived s*tate-of-knowledge* distribution for the measured quantity gives the full information about the quantity, based on the currently available information.

In order to simplify the notation and for illustration purposes, here we consider only the simplest measurement model of a single scalar quantity, say Y, which can be expressed by a functional relationship

$$Y = f(X_1, \dots, X_n),\qquad (1.A1)$$

where Y is the scalar output quantity of interest (the measurand) which depends functionally on input quantities $X_1, \ldots, X_n$. Frequently, it is adequate to assume that the functional relationship of the considered measurement model is linear (at least approximately), of the following additive linear form

$$Y = c_1 X_1 + \cdots + c_n X_n, \qquad (1.A2)$$

where the input quantities $X_1, \ldots, X_n$ are mutually independent random variables with known probability distributions. Here, $c_1, \ldots, c_n$ denote the known constants and Y represents the univariate output quantity (a random variable with an unknown distribution to be determined).

Under such assumptions it is possible to derive theoretically the exact convolution-type distribution of the output quantity Y, but analytical derivation of such distribution is typically too complicated. If this is the case, the possible alternative approach should be based on numerical (computer intensive) methods and/or other approximate approaches for uncertainty evaluation as suggested in GUM and its Supplements, [2], [3], and [4]. However, assuming linearity of the measurement model and stochastic independence of the input quantities is rather strong assumption, which may be inadequate in many important situations, and usage of the nonlinear functional relation in measurement model (1.A1) and the joint multivariate distribution of $X = (X_1, \ldots, X_n)$ becomes an indispensable requirement for correct uncertainty analysis.

This is exactly the situation for which application of the approximate numerical Monte Carlo methods was suggested in the GUM Supplements [3] and [4]. Then, such derived (approximate) state-of-knowledge distribution of the output quantity is used to properly express uncertainty in measurement, and to provide coverage intervals for the stipulated coverage probability, for that quantity.

In general, the *GUM uncertainty framework* (GUF) provides a method for assessing uncertainty based on the law of propagation of uncertainty and the characterization of the output quantity by a Gaussian distribution or a scaled and shifted *t*-distribution. Supplement 1 is concerned with the propagation of probability distributions through a measurement model as a basis for the evaluation of measurement uncertainty, and its implementation by a *Monte Carlo method* (MCM). Supplement 2 describes a generalization of MCM to obtain a discrete representation of the joint probability distribution for the output quantities of a multivariate model. An alternative tool to form the state-of-knowledge probability distribution of the (scalar) output quantity in linear measurement model, based on the numerical inversion of its characteristic function, which is defined as a Fourier transform of its probability density function (PDF) is the *characteristic functions approach* (CFA).

1.A.2. GUM Uncertainty Framework (GUF)

GUF is applicable when the input quantities are summarized in terms of *best estimates* (based on the measured values and expert knowledge) and *standard uncertainties* associated with these estimates and, when appropriate, covariance associated with pairs of these estimates.

In fact, GUF provides a framework for assessing uncertainty conditionally, based on knowledge from the stated statistical moments (expectations and covariances) of the input random variables X_i only, i.e. without full knowledge about the distribution of the input quantities, and further by using the law of propagation of uncertainty. Then the characterization of the output quantity (expressed by the best estimate and its uncertainty and/or the expanded uncertainty) is expressed by using the Gaussian distribution or a scaled and shifted t-distribution. Such distribution is derived under quite general assumptions by using the asymptotic approximation based on the *Central Limit Theorem* or its small sample approximation by Student's t-distribution with the Satterthwaite's estimate of the associated degrees of freedom. In general, application of GUF is simple but if the explicit or implicit assumptions are not justified, it could lead to more or less incorrect results. Thus, validation of GUF by other methods is highly recommended.

1.A.3. Monte Carlo Method (MCM)

MCM is a numerical approximate method to determine the state-of-knowledge distribution of the output quantity Y, defined by the used measurement model (A1), in situations where the conditions for the GUF are not fulfilled, or it is unclear whether they are fulfilled. MCM is based on the following steps:

(i)   Assignment of the (joint) probability distribution to the input quantity $X = (X_1, \ldots, X_n)$ on the basis of all currently available knowledge;

(ii) Determination of an empirical discrete representation of the probability distribution for the output quantity, based on realization of a large number of artificial Monte Carlo experiments propagating the possible realizations of X (probability distribution for X) through the model to the possible realizations of Y (probability distribution for Y); and based on that,

(iii) Representation of the best estimate of the output quantity and the evaluation of the associated standard uncertainty. Principal advantage of MCM is its simplicity and consistency (convergence to the true distribution with growing number of the Monte Carlo simulations). A disadvantage of MCM is a large demand on the computational resources and inherent non-uniqueness of the results. Often, to achieve the pre-selected accuracy a need of a very large number of simulations is required.

1.A.4. Characteristic Functions Approach (CFA)

An alternative to MCM is CFA suggested in [6] to form the state-of-knowledge probability distribution of the output quantity in linear measurement model, based on the numerical inversion of its characteristic function (CF), defined as a Fourier transform of its PDF.

Table 1.3 presents CFs of selected univariate distributions frequently used in metrological applications for modeling marginal distributions of the input quantities. Notice that CFs of the symmetric zero-mean distributions are real functions of its argument t, while in general, CFs are complex valued functions.

CF of a weighted sum of independent random variable is simple to derive if the measurement model (1.A1) is linear and the input quantities X_i are independent. In such situations, CF of the output quantity Y, defined by (1.A2), is given as

$$\mathrm{cf}_Y(t) = \mathrm{cf}_{X_1}(c_1 t) \times \cdots \times \mathrm{cf}_{X_n}(c_n t), \qquad (1.A3)$$

where $cf_{X_j}(t)$ denote the known CF of X_j. Then, the distribution of the output quantity Y, the probability density function and the cumulative distribution function (CDF), can be derived by numerical inversion of its CF, which can be efficiently calculated by a simple trapezoidal quadrature:

$$pdf_Y(y) \approx \frac{\delta_t}{\pi} \sum_{j=0}^{N} w_j \, \Re\left(e^{-it_j y} cf_Y(t_j)\right)$$

$$cdf_Y(y) \approx \frac{1}{2} - \frac{\delta_t}{\pi} \sum_{j=0}^{N} w_j \, \Im\left(\frac{e^{-it_j y} cf_Y(t_j)}{t_j}\right), \qquad (1.A4)$$

where N is the sufficiently large integer, w_j are the trapezoidal quadrature weights, with $w_0 = w_N = 0.5$ and $w_j = 1$ for $j = 1, \dots, N-1$, t_j are equidistant nodes from the interval $(0, T)$, for sufficiently large T, and $\delta_t = \frac{T}{N}$. By $\Re(z)$ and $\Im(z)$ we denote real and imaginary part of the complex value z, respectively.

Table 1.3. Characteristic functions of continuous univariate distributions used in metrological applications (selected symmetric zero-mean distributions). Here, $J_\nu(z)$ is the Bessel function of the first kind and $K_\nu(z)$ denotes the modified Bessel function of the second kind.

Probability distribution	Characteristic function (CF)
Gaussian, $N(0,1)$	$cf(t) = e^{-\frac{1}{2}t^2}$
Student's t, t_ν	$cf(t) = \frac{1}{2^{\frac{\nu}{2}-1}\Gamma\left(\frac{\nu}{2}\right)} \left(\nu^{\frac{1}{2}}|t|\right)^{\frac{\nu}{2}} K_{\frac{\nu}{2}}\left(\nu^{\frac{1}{2}}|t|\right)$
Rectangular, $R(-1,1)$	$cf(t) = \frac{\sin(t)}{t}$
Triangular, $T(-1,1)$	$cf(t) = \frac{2 - 2\cos(t)}{t^2}$
Arcsine, $U(-1,1)$	$cf(t) = J_0(t)$

The algorithms for computing the approximations suggested in (A4) have been implemented as a part of the *CharFunTool: The characteristic functions toolbox in MATLAB. CharFunTool* consists of a set of algorithms for evaluating selected characteristic functions and algorithms for numerical inversion of the combined and/or compound characteristic functions, used to evaluate CDF, PDF, and/or the quantile function. The toolbox comprises different inversion algorithms, including those based on simple trapezoidal quadrature rule for computing the integrals defined by the Gil-Pelaez formulae, and/or based on using the FFT algorithm for computing the Fourier transform integrals.

For more details see [6], [7]. The characteristic functions toolbox is available at the webpage: https://github.com/witkovsky/CharFunTool.

Chapter 2

Fundamental Principles of Spectral Methods Related to Discrete Data

Martin Seilmayer & Matthias Ratajczak

2.1. Introduction

Until the mid 20th century, the groundbreaking works of Claude Shannon regarding information theory laid the foundation for the vast field of digital signal processing and spectral data analysis, and thereby enabled almost all modern daily life information technologies we are now used to. Digital telecommunication, audio and video compression are just some examples of advanced signal processing which are not possible without a theory about discrete data and its spectral representation. Signal processing starts with the process of sampling by means of a mathematical model, which results in a mapping between the continuous physical measure – like temperature, voltage fluctuations of a microphone or a camera picture – and its time and value discrete representation. According to that, information about the signal can be gained by filtering, data manipulation, pattern extraction and related procedures. Because of the vast variety of applications, many different spectral methods were developed which utilize different mathematical transforms, for example the Fourier transform or the Hilbert transform. Utilizing the appropriate transform for a problem also opens up shortcuts in calculations, or makes signal features visible by decomposing the signal into the associated spectral domain.

Martin Seilmayer
Institute of Fluid Dynamics, Helmholtz-Zentrum Dresden-Rossendorf, Dresden, Germany

The following chapter aims at giving a brief, yet mostly complete instructive overview about the fundamental basics of sampling theory and its application in terms of Fourier transform and other in their continuous and discrete version. The chapter is split into 3 successive sections, which point at different levels of abstraction and complexity.

Section 2.2 focuses on fundamental signal processing methods and provides a concise reference for the most relevant mathematical formulations. The following Section 2.3 expands this overview by the necessary mathematical background, which must be properly considered by the users. In the course of this, typical questions like the origin of frequency or band limitation of signals and the source of artifacts will be discussed. The subsequent Section 2.4 is concerned with common applications of the most important one-dimensional methods, their properties, advantages and drawbacks as well as mistakes the user should avoid. Corresponding derivations and suggestions for further reading will be given where needed. One of the last given applications will be time-dependent spectral analysis, with which one can evaluate the point in time when a certain frequency appears in the signal of interest. A special topic of the presented methodology is the processing of fragmented data, which becomes relevant in the last subsection. Here, the Lomb-Scargle method will be explained with an illustrative example to deal with this special type of signal.

Individual application examples are supported by source code programmed in the statistical language R [1] with the help of M. Seilmayer's *spectral* [2] package. The main goal of the provided R code and introduced package is an easy-to-use user interface, providing naturally scaled output data according to the input. This means if the input time vector would be in units of seconds the output of the spec.fft(x,y)-function would be the natural frequency unit Hz. Besides the analytical theory, the application parts of each subsection will show how spectral methods can be utilized to carry out or accelerate several calculations on discrete data, like derivation, integration or convolution (filtering). Here, source code snippets of the *spectral* package highlight the implementation and underline the methodology.

The following section extends the publication 'A Guide on Spectral Methods Applied to Discrete Data in One Dimension' by M. Seilmayer and M. Ratajczak [3].

2.2. Mathematical Concepts

This section introduces the basic ideas and principles leading to the methods of spectral analysis for discrete data. Here, the focus lies on measurements and time series of physical processes. For now, utilizing the time as variable t makes it easier to understand the explained methods. Of course, t might be replaced by the space variable x, if a distribution of spatial events is of interest. In this sense space and time are meant to be the *location* where the signal is defined.

The following definitions and explanations are short and sweet. For further reading it is suggested that you refer to "The Scientist and Engineer's Guide to Digital Signal Processing" by S. Smith [4] or to "Time-Frequency Analysis" by L. Cohen [5]. These two comprehensive works elaborate the underlying mathematical framework in detail. Both books are highly recommended if theoretical approaches and mathematical subtleties are of interest. But here, practical applications to measured data and an understanding of the complex connections will be focused.

2.2.1. Signal Definition

In the context of this section a signal $s(t)$ corresponds to a physical quantity $p(t)$ and is therefore real-valued and causal. This means, that with the measurement of the process $p(t)$ the signal $s(t)$ starts to exist at a certain point in time and ends later when the measurement is finished.

With that in mind, the signal function $s(t)$ represents a slice of the length T for times $t \geq 0$. This can be properly defined as follows:

$$s(t) = \begin{cases} p(t) & 0 \leq t < T, \\ 0 & \text{elsewhere} \end{cases} \tag{2.1}$$

The simple noisy signal in Fig. 2.1 illustrates that. For this example, the underlying physical process $p(t)$ could be a temperature, which is measured in the time range between $0 \leq t < 1$. Evidently the temperature of an object exists before and after the measurement takes place, so the content of $s(t)$ only maps to a time interval of this process.

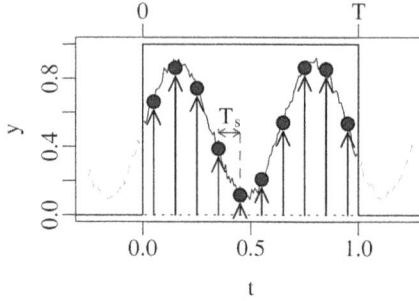

Fig. 2.1. The arbitrary signal with overlapped normal distributed noise $p(t) = 0.4\cos(2\pi \cdot 1.5t - \pi/2) + \mathcal{N}(0, 0.02) + 0.5$ is shown here in grey. The measurement takes place only within the rectangular window where the signal $s(t)$ exists, shown as black line. Outside this window the physical process might go on, but is not recognized anymore. The points and arrows mark the sampling of that signal.

With respect to a digital representation of information, the measurement of $p(t)$ takes place by acquiring data in the form of sampling the continuous signal at N different instances in time. The resulting sampled signal series

$$\hat{s}_n = \sum_{n=0}^{N-1} s(n \cdot T_s) \cdot \vec{e}_n \qquad (2.2)$$

is the data vector $\hat{s}_n$. Here, the index n addresses the nth element in this N-dimensional data vector $\hat{s}$. In Eq. (2.2) the basis vector $\vec{e}_n$ ensures that $\hat{s}$ becomes a vector in a mathematical sense, which simply represents an ordered list of elements. Remember, the absolute value $|\vec{e}_n| = 1$ is always equal to one, so the sum only helps to iterate along the continuous signal $s(t)$.

2.2.2. The Fundamental Idea of Frequency Analysis

To explain the idea of frequency (or spectral) analysis right from the beginning, it is necessary to define the signal function $s(t)$. From now on, the signal represents a sum of sinusoidal functions in the form

$$s(t) = \sum_i A_i \cdot \cos(\underbrace{2\pi f_i t}_{\omega_i} + \varphi_i), \qquad (2.3)$$

which, in the final consequence, leads to the expansion into a model consisting of a set of orthogonal functions.

The aim of frequency analysis is to calculate the amplitude and phase spectrum of a signal. The result of this procedure is a mapping of signal properties in the time domain to their corresponding frequencies. With the assumption that a measured time series of a physical process is the sum of many individual sinusoidal functions (see Eq. (2.1)), it could be of interest to estimate each of the amplitudes A_i and phases φ_i for given frequencies f_i.

The first approach to do that would be to estimate each A_i and φ_i by a least squares fit against a sinusoidal model (which we assume for $s(t)$) for each frequency of interest. This is a legitimate attempt with certain drawbacks and some outstanding advantages, which are discussed below in Section 2.4.8.

The root of a more general method to determine the spectral components of a signal are the two common addition theorems

$$\cos\alpha \cdot \cos\beta = \frac{1}{2}\left(\cos(\alpha - \beta) + \cos(\alpha + \beta)\right) \qquad (2.4)$$

$$\cos\alpha \cdot \sin\beta = \frac{1}{2}\left(\sin(\beta - \alpha) + \sin(\alpha + \beta)\right), \qquad (2.5)$$

which mark the starting point of the following explanations. A collection of addition theorems can be found in almost any book of formulae and tables like I. Bronstein [6].

Given a signal with only one frequency $\omega = 2\pi \cdot f$ and amplitude A in the continuous time domain

$$s(t) = A \cdot \cos(\omega t + \varphi), \qquad (2.6)$$

which additionally fits precisely into the time interval

$$T = \frac{n}{f} \quad \textit{for } n \in \mathbb{N} \qquad (2.7)$$

The multiplication with $\cos(\omega_0 t)$ as well as $\sin(\omega_0 t)$ and finally integrating over all times yields

$$\int_{-\infty}^{\infty} s(t) \cdot \cos(\omega_0 t) \; dt =$$

$$= \frac{A}{2} \left(\int_{-\infty}^{\infty} \cos\big((\omega - \omega_0)t + \varphi\big) \; dt + \underbrace{\int_{-\infty}^{\infty} \cos\big((\omega + \omega_0)t + \varphi\big) \; dt}_{=0} \right) \quad (2.8)$$

$$= \begin{cases} \dfrac{A}{2}\cos(\varphi)\displaystyle\int_0^T dt = \dfrac{A}{2}\cos(\varphi)T &, \omega = \omega_0 \\[2mm] \to 0 &, \omega \neq \omega_0 \end{cases} \quad (2.9)$$

$$= R \quad \text{(real part or in phase component)} \quad (2.10)$$

for the cos-term, which is going to be named R. Herein, the variable ω_0 denotes the frequency of interest. The sin-term is then very similar

$$\int_{-\infty}^{\infty} s(t) \cdot \sin(\omega_0 t) \; dt = \begin{cases} \dfrac{A}{2}\sin(\varphi)T &, \omega = \omega_0 \\[2mm] \to 0 &, \omega \neq \omega_0 \end{cases} \quad (2.11)$$

$$= I \quad \text{(imaginary part or} \\ \text{quadrature component)} \quad (2.12)$$

and is called I. In both equations on the right hand side the right integral vanishes to zero, because each positive area element of a sin-function can be mapped to a negative area element. Note, only in the case of $\omega_0 = \omega$ a phase-dependent part is left. Since the signal $s(t)$ is defined as a bounded function in an interval $0 \leq t < T$, the integration boundaries can be reduced to $\int_0^T dt = T$. With this in mind, the calculation of amplitude and phase can be performed, as follows.

With the trigonometric identity $\sin^2 \alpha + \cos^2 \alpha = 1$ the amplitude can be calculated as

$$A = \frac{2}{T}\sqrt{R^2 + I^2}, \quad (2.13)$$

where R and I are *non-normalized* components, see Eqs. (2.9) and (2.11). The phase information is then gained by

$$\varphi = \arctan\left(\frac{I}{R}\right) \qquad (2.14)$$

accordingly. The introduced procedure is called quadrature demodulation technique (QDT), which describes amplitude demodulation in the field of transmission theory. Anyway, by evaluating the amplitude A at many different frequencies ω_0, a periodogram emerges, which maps amplitudes in frequency space to sinusoidal processes in time space.

Note: To clarify the relation to the mathematical background, the normalization of A to $2T^{-1}$ is carried out in Eq. (2.13). In practice R and I are normalized by T or 2π (for generalized expressions) right from the beginning (Eq. (2.8)) so they can be interpreted as the real and imaginary part of a signal decomposition or as Fourier coefficients. The complex exponential formulation then corresponds to

$$s(t) = \Re\left(A \cdot e^{i\omega t + \varphi_i}\right) \qquad (2.15)$$

with the Euler identity $e^{i\phi} = \cos\phi + i\sin\phi$.

Until now it is assumed that the signals period fits perfectly (as an integer value) in the time range $0 \leq t < T$. If this is not the case, one of the main disadvantages arises. The method produces a truncation error ε, which depends on the remainder after the modulus division

$$n = (2\pi f \cdot T)\mathrm{mod}(2\pi) \qquad (2.16)$$

with respect to the total time interval T. In other words, if the period $2\pi / \omega$ of the signal does not fit completely n times into the integration range, the integrals containing $\omega + \omega_0$ will not vanish completely. However, the right integral term of equation (2.8) can be split into an $n \cdot 2\pi$ periodic part and a residue. Then, the periodic integral of $\cos\big((\omega + \omega_0)t + \varphi\big)$ vanishes,

$$R = \frac{A}{2}\left(\int_0^T \cos\big((\omega - \omega_0)t + \varphi\big)\ dt + \int_0^T \cos\big((\omega + \omega_0)t + \varphi\big)\ dt\right)\Bigg|_{\omega = \omega_0} =$$

$$= \frac{A}{2}\left(T\cdot\cos(\varphi) + \underbrace{\int_{0}^{\frac{n\cdot2\pi-\varphi}{\omega+\omega_0}} \cos\big((\omega+\omega_0)t+\varphi\big)\ dt}_{\rightarrow 0} + \underbrace{\int_{\frac{n\cdot2\pi-\varphi}{\omega+\omega_0}}^{T} \cos\big((\omega+\omega_0)t+\varphi\big) dt}_{=\varepsilon} \right) =$$

$$= \frac{A}{2}\big(T\cdot\cos(\varphi)+\varepsilon\big), \tag{2.17}$$

whereas a residue ε remains in the second integral from the last $n\cdot2\pi$ period up to the full length T. Obviously, a minimum of ε is reached if $T=n\cdot2\pi$. All this holds if $\omega=\omega_0$. If this is not the case, also the integral $\int\cos\big((\omega-\omega_0)t+\varphi\big)$ splits into two parts from 0 to $n\cdot2\pi$ and a residue, which leads to a second error ε_2. Finally, with $\omega\approx\omega_0$ this leakage effect will produce artificial amplitudes in the neighborhood of ω_0. Anyway, by inserting the result (2.17) in Eq. (2.13) or (2.14), mixed products arise, which makes it difficult or even impossible to reject the error ε. All this is far away from being mathematically complete, but it points in the direction towards an error discussion, which is presented in more detail by A. Jerri [7, Chap. 6].

At this point, the previously mentioned statistical approach "fitting a sinusoidal model to the data set" becomes attractive. The nature of a fitting algorithm enables the minimization of the phase error $\varepsilon\rightarrow0$ which leads to an optimal result. A brief discussion of this concept is given in "Studies in astronomical time series analysis. II" by J. Scargle[8]. Section 2.4.8 provides an introduction to the Lomb-Scargle method, which is somewhat equal to a least square fit of a sinusoidal model to the given data. Section 2.4.8 provides an overview of this method.

2.2.3. Integral Transforms

In contrast to the previously described QDT, which is more or less a statistical approach, the integral transforms are the mathematical backbone of spectral analysis. An integral transform represents the bijective mapping $s(t)\leftrightarrow S(\omega)$ between time and frequency domains, which states that all information present in the first will be transformed to the second domain. In the following, capital letters always denote the frequency domain, whereas lowercase letters represent the time (or spatial) domain of the corresponding variable.

Fourier transform. First of all the Fourier transform (FT) should be introduced. Here, the definitions and some of the properties in the continuous and discrete time domain are different. A comparison is given in Table 2.1. By means of FT, the signal $s(t)$ forms a Fourier pair with $S(\omega)$ and has to fulfill some requirements. According to the definition given in Table 2.1 the Fourier transform can be written as

$$S(\omega) = \int_{-\infty}^{\infty} s(t) \cdot e^{-i\omega t} \, dt = F\ (s(t)), \qquad (2.18)$$

which utilizes the notation with the Fourier operator F. Next to that, the inverse back transform is denoted by F^{-1}.

Table 2.1. Continuous and discrete Fourier transform.

	Continuous	Discrete
Transform	$S(\omega) = \int_{-\infty}^{\infty} s(t) \cdot e^{-i\omega t} \, dt$	$\hat{S}(m) = \dfrac{1}{N} \sum_{n=0}^{N-1} \hat{s}(n) e^{-im\frac{2\pi}{N}n}$
Back transform	$s(t) = \dfrac{1}{2\pi} \int_{-\infty}^{\infty} S(\omega) \cdot e^{i\omega t} \, d\omega$	$\hat{s}(n) = \sum_{m=0}^{N-1} \hat{S}(m) e^{im\frac{2\pi}{N}n}$

Now the question arises: Does the FT exist for all possible real-valued $s(t)$? The answer is no, because there are a number of conditions the signal must not violate in order to proceed with FT and related signal analysis. First, let's estimate the absolute maximum value of $s(t)$

$$|S(\omega)| \le \int_{-\infty}^{\infty} |s(t) \cdot e^{-i\omega t}| \, dt$$

$$\le \int_{-\infty}^{\infty} |s(t)| \cdot \underbrace{|e^{-i\omega t}|}_{=1} \, dt \qquad (2.19)$$

$$\le \int_{-\infty}^{\infty} |s(t)| \, dt < \infty, \qquad (2.20)$$

which should result in a finite value to gain a reasonable statement about $s(t)$. Note, the rotating pointer $\left|e^{i\omega}\right| = 1$ has unity length. Eq. (2.20) can be extended to signals which exist in a time interval $T = t_2 - t_1$. Starting from (2.19), and denoting the maximum value of $|s(t)|$ with s_{max}

$$\left|S(\omega)\right| \leq \int_{t_1}^{t_2} |s(t)| \cdot \left|e^{-i\omega t}\right| \, dt$$

$$\leq \int_{t_1}^{t_2} |s(t)| \, dt \leq \int_{t_1}^{t_2} s_{max} \, dt$$

$$\leq s_{max} \cdot (t_2 - t_1) < \infty \qquad (2.21)$$

it can be concluded that a time-limited signal $s(t)$ must not contain any poles (or any other infinite values) in order not to violate condition (2.21). At last, causal signals $s(t)$ with $t \geq 0$ might be of interest. Here, an additional restriction applies, so that the spectrum

$$\left|S(\omega)\right| \leq \int_0^\infty |s(t)| \cdot \left|e^{-i\omega t}\right| \, dt \quad \text{with} \, |s(t)| \leq M \cdot e^{pt}$$

$$\leq \int_0^\infty M \cdot e^{pt} < \infty \qquad (2.22)$$

only exists if $p < 0$. From this follows that a causal signal $s(t)$ should be convergent with $\lim_{t \to \infty} s(t) = 0$.

The Fourier transform provides different features like frequency shifting

$$\mathsf{F}\left(e^{ibt} f(t)\right) = F(\omega - b) \qquad b \in \mathbb{R}, \qquad (2.23)$$

which becomes important when amplitude modulation in Section 2.4.4 is discussed.

In addition to the characteristics of the continuous Fourier transform, the discrete Fourier transform (DFT) has a 2π- or f_s-periodic spectrum. As shown in Section 2.3.1, the result of the DFT is a complex valued data vector, in which the first element $\hat{S}_1$ *always* contains the mean value of the data series. Subsequent bins hold the amplitude values up to $\hat{S}_{N/2}$.

From this position on the spectrum is mirrored and repeats up to the sampling frequency. In Section 2.3 the symmetry of the DFT's and its consequences are discussed in more detail. It will be shown how to take advantage of this, e. g. to handle negative frequencies. In case of one dimensional data, these negative frequencies are just a mirror symmetric copy of their positive counterparts, but in case of more dimensional (multivariate) data, the quadrant (sign of frequency vector) of the coordinate system where a significant peak is located contains information about sign, direction and drift speed of the corresponding wave in n-dimensional spatial (time) domain.

Properties of the Fourier operator. In the following, the properties of the Fourier operators [1] are listed.

Linearity.

$$\mathsf{F}\ \big(a \cdot f(t) + b \cdot g(t)\big) = a \cdot F(\omega) + b \cdot G(\omega) \tag{2.24}$$

Scaling.

$$\mathsf{F}\ \big(f(t/a)\big) = |a| \cdot F(a \cdot \omega) \qquad a \in \mathbb{R} \text{ and } a \neq 0 \tag{2.25}$$

Translation.

$$\mathsf{F}\ \big(f(a \cdot t + b)\big) = \frac{1}{a} e^{i\omega \cdot b/a} F(\omega/a) \tag{2.26}$$

$$\text{with } a, b \in \mathbb{R} \text{ and } a \neq 0 \tag{2.27}$$

Damping.

$$\mathsf{F}\ \big(e^{ib \cdot t} f(a \cdot t)\big) = \frac{1}{a} F\big((\omega - b)/a\big) \qquad a \in \mathbb{R}^{+} \tag{2.28}$$

n^{th} Derivation. The function $f(t)$ must be absolutely integrable and $\lim_{t \to \pm\infty} f(t) = 0$

$$\mathsf{F}\ \left(\frac{d^{(n)}}{dt^{(n)}} f(t)\right) = i\omega^{(n)} \mathsf{F}\ \big(f(t)\big) \tag{2.29}$$

Integration. Only if $\int_{-\infty}^{\infty} f(t)\, dt = 0$ then

$$\mathsf{F}\left(\int_{-\infty}^{t} f(t)\, dt\right) = \frac{1}{i\omega} F(\omega) \qquad (2.30)$$

Convolution. Only if the integrals $\int_{-\infty}^{\infty} |f(t)|^2\, dt$ and $\int_{-\infty}^{\infty} |g(t)|^2\, dt$ are existent, the two sided convolution of $f(t)$ and $g(t)$

$$f(t)*g(t) = \int_{-\infty}^{\infty} f(\tau)g(t-\tau)\, d\tau \qquad (2.31)$$

can be calculated with the help of the Fourier transform. In the frequency domain the convolution operation reduces to a multiplication. This is expressed as follows:

$$\mathsf{F}\ (f(t)*g(t)) = \mathsf{F}\ (f(t)) \cdot \mathsf{F}\ (g(t)) \qquad (2.32)$$

This statement also holds for the opposite operation: A convolution in the frequency domain corresponds to a multiplication in the time domain:

$$\mathsf{F}\ (f(t)) * \mathsf{F}\ (g(t)) = f(t) \cdot g(t) \qquad (2.33)$$

Hilbert transform. Next to the Fourier transform the Hilbert transform (HT) might be defined as real-valued convolution operation on a real-valued signal $s(t)$

$$\mathsf{H}\ (s(t)) = \frac{1}{\pi} \int_{-\infty}^{\infty} \frac{s(t')}{t-t'}\, dt' \qquad (2.34)$$

Additionally, the linear operator H is introduced to execute the Hilbert transform on a function or data set. From the representation in the Fourier space the constant phase shifting feature becomes clear:

$$\mathsf{F}\ (\mathsf{H}\ (s(t))) = -i \cdot \text{sign}(\omega) \cdot \mathsf{F}\ (s(t)) \qquad (2.35)$$

Hence, Eq. (2.35) can be used to easily calculate $\mathsf{H}\ (s(t))$ in terms of a Fourier transform. Remember, here the $\text{sign}(\omega)$ function is defined on $-\infty < \omega < \infty$, which must be taken into account when performing $\mathsf{H}\ (\hat{s}(n))$ on discrete data. In this case, the discrete Hilbert transform is calculated by

$$\mathsf{F}\left(\mathsf{H}\left(\hat{s}_n\right)\right) = \begin{cases} \mathrm{i}\cdot\mathsf{F}\left(\hat{s}_n\right) & 0 \leq n < N/2 \\ -\mathrm{i}\cdot\mathsf{F}\left(\hat{s}_n\right) & N/2 \leq n < N \end{cases} \qquad (2.36)$$

This formalism considers the fact that "negative" frequencies are mirrored into the upper parts of the DFT data vector. However, the result of $\mathsf{H}\left(s(t)\right)$ is real-valued, with each frequency component phase-shifted by $\pm\pi$. In conjunction with a real-valued and causal signal $s(t)$, the HT helps to represent the real physical character of measured data, which should only contain positive frequencies, due to real worlds processes. From this, the analytical signal

$$a(t) = s(t) + \mathrm{i}\mathsf{H}\left(s(t)\right) \qquad (2.37)$$

follows, which is a complex representation of $s(t)$ containing a one-sided Fourier spectrum.

One derivation of the Hilbert transform in conjunction with the analytical signal is given in Section 2.3.3 and its application in terms of data filtering is briefly discussed in Section 2.4.3 and the following.

In addition to that, the concept of instantaneous frequency can be explained with the help of the HT, which is also demonstrated by the example in Section 2.4.3. This topic is discussed in detail in "The Empirical Mode Decomposition and the Hilbert Spectrum for Nonlinear and Non-Stationary Time Series Analysis" by N. Huang [9]. For further reading on that "Amplitude, phase, frequency – fundamental concepts of oscillation theory" by D. Vakman [10] should be mentioned.

2.3. Background and Subtleties of Spectral Methods with Discrete Data

After the first section gave an overview about the most important concepts about spectral analysis of discrete data, the descriptions are to be expanded by the most relevant mathematical derivations. This part of the section is therefore concerned with the background of the discrete Fourier transform (DFT), the Nyquist condition and the Hilbert transform. These will provide the reader with a deeper understanding of the methods in the subsequent application section.

The following formulations are being used in this section. First of all, let the signal function

$$s(t) = \sum_{i=0}^{m} A_i \cdot \cos(\omega_i t + \varphi_i) + \mathsf{N}\,(0, \sigma) \qquad (2.38)$$

be a set of sinusoidal functions plus an optional noise term. This signal is then sampled with an equidistant spacing T_s, so the discrete data vector

$$\hat{s}_n = \sum_{n=0}^{N-1} s(n \cdot T_s) \cdot \vec{e}_n$$

of length N represents the sampling series. Using the discrete Fourier transform, according to Table 2.1, the spectrum of the data vector $\hat{s}$

$$\hat{S}_m = \mathsf{F}\,(\hat{s}_n)$$

can be calculated. In the following this set of definitions will be the starting point to describe different methods and corresponding subtleties to pay attention on.

A deep and profound introduction can be found in "The Shannon sampling theorem— Its various extensions and applications: A tutorial review" by A. Jerri [7]. A detailed comprehensive mathematical work is given in the book 'Fourier Analysis' by Z. Duoandikoetxea [11].

2.3.1. Derivation of the Discrete Fourier-Transform

First of all, the derivation of discrete Fourier transform will be carried out to clarify its properties and conditions. It will be shown that the spectral information becomes periodic and in the 1D case redundant.

Let y be a real-valued function of time, with

$$y = s(t) : t \in \mathbb{R} \rightarrow s(t) \in \mathbb{R}, \qquad (2.39)$$

which is going to be sampled later on. Fig. 2.2 illustrates the continuous function and its sampling points. One possible way to perform the sampling utilizes the Dirac-function as a distribution as follows:

$$\delta(t) = \begin{cases} \infty, & t = 0 \\ 0, & t \neq 0 \end{cases} \tag{2.40}$$

$$\text{with } \int_{-\infty}^{\infty} \delta(t) \, dt = 1 \tag{2.41}$$

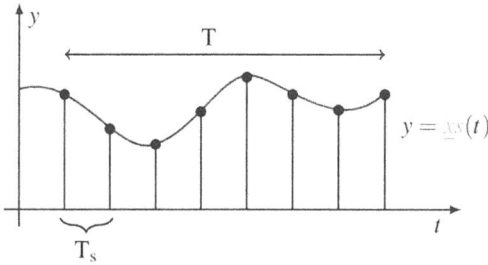

Fig. 2.2. Common band-limited and real-valued function. The reference sampling points are equally spaced with the distance T_s.

The Dirac pulse is described as $\delta(t \neq 0) = 0$ for all values of t except $t = 0$. At this point the Dirac function returns an infinite value, $\delta(t = 0) \to \infty$. However, the behavior of (2.41) turns the Dirac function into a distribution, which enables the conservation of signal energy when integrating along a sample. The finite value of this integral enables subsequent calculations to sample a point from the continuous function $y = s(t)$. This is called "sifting" property and can be written as

$$\int_{-\infty}^{\infty} \delta(t)\phi(t) \, dt = \phi(0) \tag{2.42}$$

Eq. (2.42) describes the process of sampling in a way that – in this special case – the integral of continuous function $\phi(t = 0)$ is determined at only one point. In conjunction with the Fourier transform (2.42) this leads to a necessary simplification for the derivation of the DFT.

This can be utilized to model the sampling of the continuous function $s(t)$ with the help of the sampling function $\sum_{n=-\infty}^{\infty} \delta(t - nT_s)$, which is a sum of individual Dirac pulses in a distance of $\Delta t = T_s$. The resulting sampling series

$$\dot{s}(t) = s(t)\sum_{-\infty}^{\infty}\delta(t - nT_{\mathrm{s}}) = \sum_{-\infty}^{\infty}s(nT_{\mathrm{s}})\delta(t - nT_{\mathrm{s}}) \qquad (2.43)$$

evaluates $s(t)$ at n reference points in time by multiplying the sampling function to $s(t)$. The time span T_{s} denotes the time between two successive samples. Note, because of the properties of the Dirac pulse the sum (2.43) converges to the value $\dot{s}(t) \to \infty$.

In the next step, the sampling series $\dot{s}(t)$ is going to be inserted in the Fourier integral

$$S(\omega) = \int_{-\infty}^{\infty}s(t)e^{-i\omega \cdot t}\ dt, \qquad (2.44)$$

which transforms the function $s(t)$ from the time or spatial domain into the frequency domain, expressed by the complex valued spectrum $S(\omega)$. The capital letter $S(\omega)$ supports the result, which is a bijective mapping between $s(t) \leftrightarrow S(\omega)$. At the moment, both functions $s(t)$ and its spectrum $S(\omega)$ are still continuous. For real signals $s(t)$ is required to be causal, real-valued and finite in time, with

$$0 \le t < T \qquad (2.45)$$

This means the signal is defined in a certain range of non-zero t and stops existing at a maximum time T. With this behavior the integration boundaries in (2.44) change accordingly. With respect to our model function (2.38) it is sufficient to determine the amplitudes at m individual frequencies. The result is the discrete spectrum of the Fourier series

$$S(m) = \frac{1}{T}\int_{0}^{T}\dot{s}(t)e^{-im \cdot \omega_0 \cdot t}\ dt \qquad (2.46)$$

with $m \in \mathbb{N}$. By inserting the sampling series $\dot{s}(t)$ in (2.46), now the continuous function $s(t)$ gets sampled at discrete points at $t = n \cdot T_{\mathrm{s}}$, which modifies the discrete spectrum:

$$S(m) = \frac{1}{T}\int_{0}^{T}\underbrace{\sum_{n=-\infty}^{\infty}s(nT_{\mathrm{s}})\delta(t - nT_{\mathrm{s}})e^{-im \cdot \omega_0 \cdot t}}_{\dot{s}(t)}\ dt \qquad (2.47)$$

The distribution character (2.41) and the sifting property of the Dirac function help to rearrange the equation into the DFT form [6, 12],

$$S(m) = \frac{1}{T} \sum_{n=-\infty}^{\infty} s(nT_s) e^{-im \cdot \omega_0 \cdot nT_s} \tag{2.48}$$

Finally, for a finite data set of N equidistant data points, $\omega_0 = 2\pi / T$ and a total length of $T = N \cdot T_s$, this can be rewritten as:

$$S(m) = \frac{1}{NT_s} \sum_{n=0}^{N-1} s(nT_s) e^{-im \cdot 2\pi \frac{nT_s}{N \cdot T_s}}$$

$$= \frac{1}{N} \sum_{n=0}^{N-1} s(n) e^{-im \cdot 2\pi \cdot \frac{n}{N}} \tag{2.49}$$

Here $m \in \mathbb{N}$ measures the normalized discrete frequencies on which the spectrum is being evaluated.

This discrete form of the Fourier transform provides several properties which define the spectrum $S(m)$.

1. The spectrum is periodic with respect to 2π, as indicated in Fig. 2.3. It follows that at least $m = N$ frequencies are necessary to describe a unique mapping of $s(n) \leftrightarrow S(m)$.

2. $S(\omega)$ can be interpreted as a line spectrum because it is only evaluated at $\omega = \dfrac{2\pi}{N \cdot T_s} m$ discrete frequencies.

3. A non-periodic function is set to T periodicity by definition (2.38) and the periodic spectrum (2.49).

The main feature, which arises from (2.49), is the 2π-periodic spectrum, which is the consequence of the behavior of the circular pointer $e^{i\varphi}$ described by its complex exponent $-i \cdot 2\pi \cdot n \cdot m / N$. Regarding the two-sided infinite integration boundaries of (2.44) and the following steps, it becomes clear that $X(0...N)$ represents a unique spectral pattern of finite length, as long as the signal contains sufficiently few spectral components, which fit into the range of $m = 0...N/2$ or $0...\pi$

respectively. Since the spectrum is periodic, the normalized frequency range $\pi \ldots 2\pi$ equals the range $-\pi \ldots 0$, and therefore both formulations are equivalent at first glance. Depending on the problem to be solved, the proper choice of either of these formulations will facilitate computations significantly (compare the examples in Section 2.4.1 concerning oversampling and the derivative of a function). The formulation with negative frequencies is especially useful in multidimensional spectral analysis.

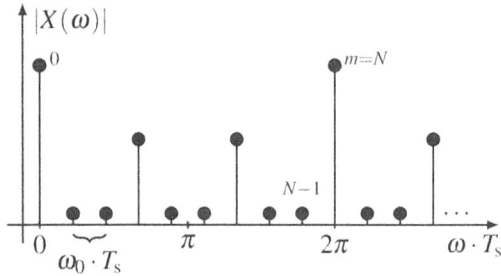

Fig. 2.3. DFT with 2π-periodic spectrum. The dimensionless frequency axis ranges from 0 to 2π whereas its resolution ω_0 is determined by the length of the total data-set $T = N \cdot T_s$. It follows that the Nyquist frequency can be found at π. In case that $m \geq N$, which is equal to $\omega \geq 2\pi / T_s$, the spectrum repeats. At $m = 0$ the plot indicates the mean value of the analyzed data. An arbitrary higher frequency component is indicated at $m = 3$.

As a consequence of the symmetry around $\omega = 0$ (see Section 2.3.2) and the periodicity, frequency components larger than half of the sampling frequency (no band limitation present) are projected into the lower half $(0 \ldots \pi)$ of the spectrum. Remember, this behavior is because of the circular pointer. In case of a non-band-limited signal, an explicit mapping between a certain spectral component and the underlying physical process becomes almost impossible, until additional information help to resolve this "undersampling" condition and enable a remapping. An illustrative example is given in the next section.

2.3.2. The Origin of $f_s/2$

The process of sampling a continuous function defines the time resolution, since the values of the function are not known at any point, but just at certain, typically equidistant sampling points. This reduction leads to the question of which spectral components of the signal can be resolved for a given sampling frequency. In order to achieve a *unique* representation of function $y(t)$ in the frequency space $Y(\omega)$, the signal is required to be band-limited.

To derive the band limitation we have to assume a function $y(t)$ for which the Fourier transform $Y(\omega)$ exists. There might also be a natural maximum frequency $\omega_{max} \leq \omega_B$ in the signal (as a natural, technical assumption), so the integration boundaries of the back transform

$$y(t) = \frac{1}{2\pi} \int_{-\omega_B}^{\omega_B} Y(\omega) \cdot e^{i\omega t} \, d\omega \qquad (2.50)$$

can be limited to $|\omega_B|$. In the next step this function is sampled at discrete time instances $t = n \cdot T_s$, with a sampling frequency $f_s = 1/T_s$. It follows

$$y(n \cdot T_s) = \frac{1}{2\pi} \int_{-\omega_B}^{\omega_B} Y(\omega) \cdot e^{i\omega \frac{n}{f_s}} \, d\omega$$

$$= \frac{1}{2\pi} \int_{-2\pi f_B}^{2\pi f_B} Y(\omega) \cdot e^{i2\pi \frac{f}{f_s} n} \, d\omega \qquad (2.51)$$

By substituting $d\omega \leftrightarrow 2\pi \, df$ the modified Eq. (2.51) can be written as

$$y(n \cdot T_s) = \int_{-f_B}^{f_B} Y(2\pi f) \cdot e^{i2\pi \frac{f}{f_s} n} \, df \qquad (2.52)$$

The pointer $e^{i\varphi}$ is periodic with respect to 2π, so the span of $\Delta f / f_s$, where $\Delta f = 2 f_B$, must not exceed unity to suppress duplication. It follows

$$\frac{\Delta f}{f_s} \leq 1$$

so that the maximum integration boundaries

$$f_B \leq 0.5 \cdot f_s \qquad (2.53)$$

can be concluded in the first instance. This corresponds to the spectral representation

$$S(\omega)\begin{cases} \neq 0 & |\omega| \leq \omega_{max} \\ = 0 & \text{elsewhere,} \end{cases} \tag{2.54}$$

which is only non-zero within the range $|\omega| \leq \omega_{max}$. Outside this interval the spectrum is equal to zero. The sampling theorem by C. Shannon [13] and the proof of J. Whittaker [14] state that in such a situation even a time-limited continuous function is completely defined by a discrete and finite set of samples, even if all the samples outside the time range are *exactly zero* (see the exemplary function in Fig. 2.1). The inverse argument relates to the question of minimal required sampling frequency (twice the "Nyquist frequency") $1/T_s = f_s > 2f_{max}$, which is necessary to conserve all (unique) information about the signal.

Thus, the necessity of band limitation demands that there should not be any frequency present in a data set higher than $f_s / 2$ to achieve a unique sampling in application. Theoretically, this means a function $s(t)$ is completely determined by its discrete spectrum $S(n \cdot 2\pi f_s)$ only if the signal is band-limited to $|\Delta f| < f_s / 2$. With respect to the discrete spectrum and under the assumption that the model function (2.38) can be applied so $s(t)$ (consisting of *individual* frequency components), also the discrete sampled signal $\hat{s}_n$ provides an unique spectral representation of its continuous counterpart $s(t)$. Now, individual peaks in $S(\omega)$ do map to individual signal features in $s(t)$. The requirement of band limitation assumes $s(t)$ to be perfectly periodic and steady within the length T, which leads to a limited set of individual frequency amplitudes in the spectrum.

In case of an unsteady or non-periodic function like $s(t) = t^2$ or a square-wave function, a spectral decomposition leads to a representation with infinite number of components in the continuous $S(\omega)$ space. However, it is possible to sample such signals and carry out the discrete Fourier transform (DFT) as well, but its result has to be handled carefully as shown in the following examples.

Due to the requirement that the *bandwidth* Δf must be less than the sampling frequency, the under-sampling of functions becomes possible. This changes the meaning of a typical Nyquist limit to a limitation of bandwidth which is necessary to obtain uniqueness. Remember, almost the same argument was carried out to show periodicity of a DFT spectrum in the end of Section 2.3.1, so both effects are related to each other.

The issue of band limitation is illustrated further in Sections 2.4.1 and 2.4.6.

2.3.3. Derivation of the Analytic Signal and the Hilbert Transform

Until now real world signals are not represented very well by the requirements of FT or DFT. For instance causality – signal acquisition takes place at $t \geq 0$ and is not defined at times $t < 0$ – is not represented properly and at least negative frequencies are difficult to discuss in 1D data. To improve the methodology and to gain some interesting features, the point of view must be changed a bit. This will end up in a different signal representation which utilizes the Hilbert transform (HT). In the following a pragmatic way to deduce the HT is presented.

Given a real-valued time-dependent signal $s(t)$, the Fourier transform can be calculated by

$$S(\omega) = \frac{1}{\sqrt{2\pi}} \int_{-\infty}^{\infty} s(t) \cdot e^{-i\omega \cdot t} \, dt \qquad (2.55)$$

The question is: How is a function $a(t)$ to be defined whose spectrum matches the spectrum of $s(t)$, but consists only of positive frequencies? In terms of a back transform this can be expressed in the form:

$$a(t) = \frac{2}{\sqrt{2\pi}} \int_{0}^{\infty} A(\omega) e^{i\omega \cdot t} \, d\omega \qquad (2.56)$$

Here $a(t)$ is the unknown signal, which provides a single-sided spectrum $A(\omega)$. To maintain the signal's energy the ordinary back transform must be multiplied by a factor of 2. Assuming the previous spectrum $S(\omega) = A(\omega)$ for $\omega \geq 0$ in conjunction with (2.56) will lead to

$$a(t) = 2 \cdot \left(\frac{1}{\sqrt{2\pi}} \right)^2 \int_{0}^{\infty} \underbrace{\left(\int_{-\infty}^{\infty} s(t') \cdot e^{-i\omega \cdot t'} dt' \right)}_{S(\omega)} e^{i\omega \cdot t} \, d\omega \qquad (2.57)$$

$$= \frac{1}{\pi} \int_{-\infty}^{\infty} \int_0^{\infty} s(t') e^{i\omega \cdot (t-t')} \, dt' \, d\omega \tag{2.58}$$

The reader may pay attention to the separate time variables t and t' which refer to two different integration instances. With the identity $\int_0^{\infty} e^{i\omega t} \, d\omega = \pi \cdot \delta(t) + \dfrac{i}{t}$ and the sifting property (2.42), $a(t)$ can be factorized so that

$$a(t) = \frac{1}{\pi} \int_{-\infty}^{\infty} s(t') \left(\pi \cdot \delta(t-t') + \frac{i}{t-t'} \right) dt'$$

$$= s(t) + i \underbrace{\frac{1}{\pi} \int_{-\infty}^{\infty} \frac{s(t')}{t-t'} dt'}_{\mathrm{H}\ (s(t'))} \tag{2.59}$$

now consists of the real-valued signal $s(t)$ and its imaginary extension in form of the Hilbert transform $\mathrm{i}\mathrm{H}\ (s(t))$. Eq. (2.59) shows the analytic signal definition for $a(t)$. The convolution

$$\mathrm{H}\ (s(t)) = \frac{1}{\pi} \int_{-\infty}^{\infty} \frac{s(t')}{t-t'} \, dt' \tag{2.60}$$

can be calculated in the frequency domain by applying the Fourier transform to the Hilbert transform [11, Chapter 3]

$$\mathrm{F}\ (\mathcal{H}(s(t))) = -i \cdot \mathrm{sign}(\omega) \cdot \mathrm{F}\ (s(t)) \tag{2.61}$$

So, finally, it becomes clear that $\mathrm{H}\ (f(t))$ represents an ideal phase shifter, becasue it turns all phases of $s(t)$ by 90 degrees.

The analytic signal and the Hilbert transform have certain properties, which are discussed in detail in Section 2.4.3 and in [5, 11].

Note: The introduced identity

$$\int_0^{\infty} e^{i\omega t} \, d\omega = \pi \cdot \delta(t) + \frac{i}{t}$$

is not so obvious to resolve. The following derivation will explain this in detail. First, the exponent of the left hand side is extended with $-\varepsilon \cdot \omega$, which is then taken in the limit to zero.

$$\int_0^\infty e^{i\omega \cdot t}\, dz = \lim_{\varepsilon \to 0} \int_0^\infty e^{i\omega \cdot t - \varepsilon \cdot \omega}\, dz$$

This is now being successively reduced:

$$\lim_{\varepsilon \to 0} \int_0^\infty e^{i\omega t - \varepsilon \cdot \omega}\, d\omega = \lim_{\varepsilon \to 0} \int_0^\infty e^{-\omega(-it+\varepsilon)}\, d\omega = \lim_{\varepsilon \to 0} \frac{-1}{-it + \varepsilon} \cdot e^{-\omega(-it+\varepsilon)}\Big|_0^\infty =$$

$$= \lim_{\varepsilon \to 0} \frac{-1}{-it + \varepsilon} \cdot (0 - 1) = \lim_{\varepsilon \to 0} \frac{1}{-it + \varepsilon} = \lim_{\varepsilon \to 0} \frac{i}{t + i\varepsilon}$$

The last term is now expanded by the conjugate denominator $t - i\varepsilon$. It follows:

$$\lim_{\varepsilon \to 0} \frac{i}{t + i\varepsilon} = \lim_{\varepsilon \to 0} \underbrace{\frac{\varepsilon}{t^2 + \varepsilon^2}}_{\to \pi \cdot \delta(t)} + \frac{it}{t^2 + \varepsilon^2} = \pi \cdot \delta(t) + \frac{i}{t}$$

The last term expresses the Poisson kernel, which is the solution of the Laplace equation in the upper half plane [15]. As an remark, the expression

$$\lim_{\varepsilon \to 0} \frac{1}{\pi} \frac{\varepsilon}{t^2 + \varepsilon^2} = \delta(t)$$

equals to the Lorentz distribution and converges to the Dirac function in the limit of $\varepsilon \to 0$.

2.4. Application Examples and Hints

In this section, eight independent examples are presented, which apply the knowledge of the first two sections to real-world examples. Simultaneously, hints are given to the reader wherever necessary.

2.4.1. The Essence of Band Limitation and the Nyquist Condition

These examples are based on the considerations in Section 2.3.2 and will demonstrate the cases of oversampling and undersampling as well as the calculation of the derivative of a function in the spectral domain. In the latter part the importance of a periodic signal is highlighted and the concept of signal windowing is applied.

Example – Oversampling and violation of Nyquist's condition. The artificial function

$$y(t) = \sin(2\pi \cdot 4t) + 0.5 \cdot \cos(2\pi \cdot 2t) + 1.5 \qquad (2.62)$$

is given in Fig. 2.4a.

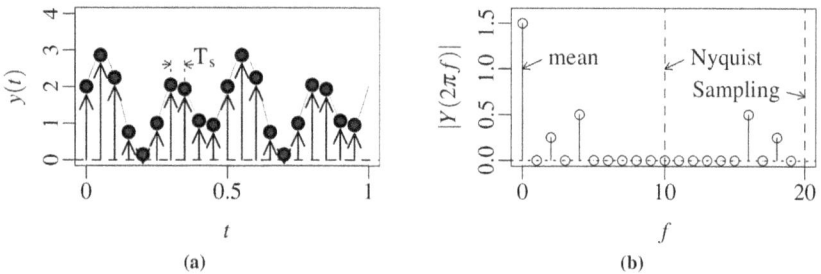

Fig. 2.4. Oversampled signal $y(t)$. Band limitation is achieved by definition of the signal. Mention that the first sample and the $i_{th} + 1$ sample at $t = 1$ would be the same, because of the periodicity of the signal.

It consists of two functions and an offset value. The sampling started at $t = 0$ and took place with a sample spacing of $T_s = 0.05$. Since $y(t)$ is perfectly periodic with $T = 1$, the first sample would be measured again after $t = T$. Since no further information can be extracted from the signal after T, sampling is stopped at this point. Because $t = T$ marks the first point of the next sampling period, it must be removed from the data set, which now has $N^* = 20$ elements in accordance with condition (2.45). Ignoring this fact would lead to an error (leakage effect) resulting in non-zero elements next to the relevant frequency peaks. The root of this behavior is discussed in Section 2.2.2 and becomes visible in the subsequent figures in this section. The resulting frequency resolution in this example is

$$\delta f = \frac{1}{T_s N^*} = 1 \qquad (2.63)$$

so, the signal's frequencies fit perfectly into this grid. Taking the DFT of the sampling series leads to the absolute amplitude spectrum $|\hat{Y}_n|$, given in Fig. 2.4b. It gains the following properties: The mean is saved in the first bin and preserves its original value of $A_0 = 1.5$, whereas the magnitude of the amplitudes at $f_{n=2} = 2$ and $f_{n=4} = 4$ are split into the upper and lower frequency domain. The reason for that can be seen in Eqs. (2.9) and (2.11), where only half the amplitude $A/2$ emerges from the calculation. To retrieve the correct signal amplitudes the individual values in both segments must be added or alternatively doubled in the range $0 \le f < f_s/2$. Remember, the DFT is mirror-symmetric to the Nyquist frequency $f_s/2 = 10$ so that the first amplitude A_1 becomes

$$A_1 = \left|\hat{Y}_{n=4}\right| + \left|\hat{Y}_{n=16}\right| \tag{2.64}$$

$$= 2 \cdot \left|\hat{Y}_{n=4}\right| \tag{2.65}$$

Fig. 2.4 and Fig. 2.5a display an oversampled signal, where the required Nyquist condition is fulfilled. The result is a unique mapping from the time domain into the frequency domain. In other words, the corresponding frequency vector $\hat{f}_n$ is fully determined in the range $0 \le n < N/2$, because it repeats inversely ordered for $N/2 \le n < N$.

However, as Fig. 2.5c indicates, if the requirement of band limitation ($f_s \ge 2f_{max}$) is violated, the mapping is not unique anymore. Remember, the frequency resolution in all given examples is $\delta f = 1$. In comparison to Fig. 2.5b, where the condition of band limitation is fulfilled, Fig. 2.5d illustrates that the upper and lower parts of the frequency bands now infiltrate each other. This makes it even more complicated to distinguish between certain frequencies and their physical correspondence. This can only be resolved if further information, like the definition of the underlying function, are available.

Example – Undersampling. The conditions of band limitation described above can be extended to the requirement

$$f_{max} - f_{min} = \Delta f \le \frac{f_s}{2},$$

which means that a unique mapping is still possible for $f > f_s$, yet only if the bandwidth Δf is not larger than the Nyquist frequency. In addition to this, the frequency range must be known. This is automatically fulfilled in the most common case, whenever the frequency ranges from $0 \le f \le f_s$.

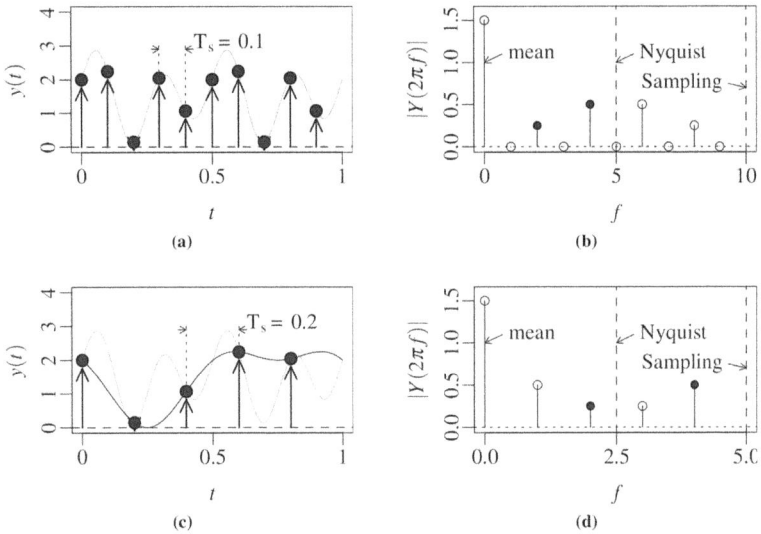

Fig. 2.5. (a, b) The signal $y(t)$ is sampled marginally. In (c, d) it is under-sampled so the upper and lower parts of the spectrum infiltrate each other. Compare the filled symbols, which correspond to the points in the lower half plane of (b). The black line in c) indicates the alternative reconstruction.

The example given in Fig. 2.6 shows the artificial function

$$y(t) = \cos(2\pi \cdot 25 \cdot t) + 0.5 \cdot \sin(2\pi \cdot 27 \cdot t) + 1.5, \qquad (2.66)$$

which is sampled with a sampling frequency of $f_s = 20$. The amplitude spectrum is given in Fig. 2.6b. Obviously $y(t)$ is under-sampled. Remember, the discrete Fourier transform calculates a periodic spectrum, which projects the high-frequency parts into the lower frequency range between 0 and f_s. Without any additional information a mapping from the spectral domain to the time domain will fail. In the present case we assume that the signal frequency is in the range $25 \le f \le 27$, so the whole spectrum can be shifted by an offset of $f_0 = f_s$, as indicated by the second f -axis. The upper and lower frequency bands can be clearly distinguished, so the mapping is unique. In contrast to that, the interpenetration of the upper and lower bands prevents such a mapping in the example in Fig. 2.5d.

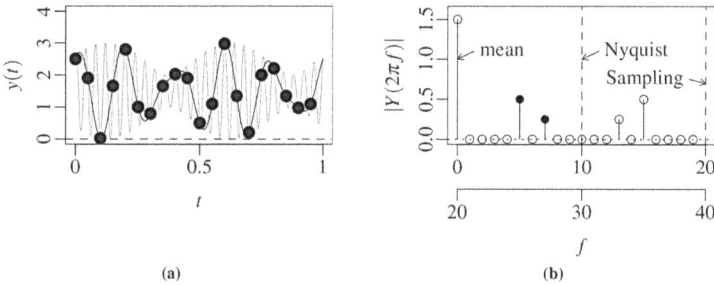

(a)   (b)

Fig. 2.6. A signal with a limited bandwidth is sampled with a lower frequency than the signal frequency. The black line in (a) displays the second possible curve, which would map to the spectrum in (b).

Example – The derivative of a function. One important property of the Fourier transform is the change of mathematical operations in the spectral domain. According to section, taking the derivative of a function reduces to a multiplication with $i\omega$

$$\mathsf{F}\left(\frac{\mathrm{d}}{\mathrm{d}x}y(x)\right) = i\omega \cdot \underbrace{\mathsf{F}\left(y(x)\right)}_{Y(\omega)} \tag{2.67}$$

in the spectral domain. In this manner, differential equations can be reduced to simpler linear algebraic representations.

In contrast to the continuous case, discrete sampled functions need to be periodic with a finite bandwidth to fulfill the requirements for the DFT. In case the function is non-periodic within the sampled window, the application of a proper window function is necessary to minimize leakage. The following code explains how to estimate the derivative of the non-band-limited function like

$$y(x) = -x^3 + 3 \cdot x \tag{2.68}$$

$$\text{with } \frac{\mathrm{d}}{\mathrm{d}x}y(x) = -3 \cdot x^2 + 3, \tag{2.69}$$

which is going to be sampled with $N = 40$ discrete points.

```
require(spectral)
# 40 sample positions
x <- seq(-2.5,2.5,length.out = 40)
# window the function and generate
# the sampling points
y <- win.tukey(x,0.2) * (-x^3+3*x)

# doing the DFT
Y <- spec.fft(y,x,center = T)
# calc deriv.
Y$A <- 1i * 2*pi * Y$fx * Y$A
# the realpart of the back transform
# contains the signal of interest
dy <- Re(spec.fft(Y,inverse = T)$y)
```

It must be mentioned that the function $y(x)$ has to be windowed in advance to achieve the periodic property. In this example this is done via the Tukey-window, which rises and falls with a quarter cycle of a cosine function (compare dashed curve). Fig. 2.7 shows the output of the listed code above. Pay special attention to the windowed version of $y(x)$ (black curve) and its almost perfect derivative. Here, the effect of the windowing process is to achieve $Y(|\omega| > 0.5 \cdot \omega_s) = 0$, which relaxes the function at the right and left side. In consequence, the resulting derivative equals the analytic solution in the mid range, but deviates at the left and right boundaries. In comparison to that, the unweighted function in (a) (gray line) produces an unusable faulty derivative in (b) (crosses). The reason for that is the discussed missing band limitation.

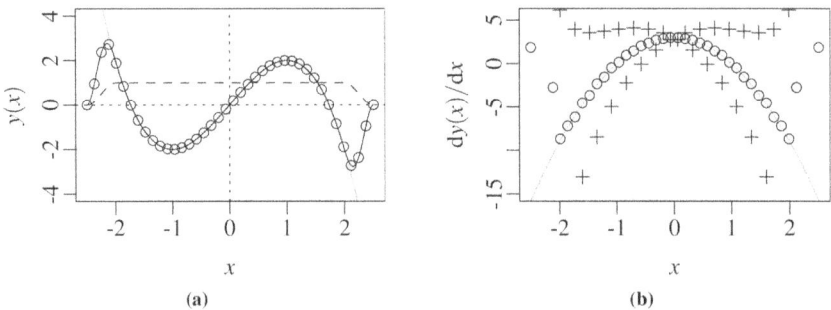

(a)   (b)

Fig. 2.7. Calculating the derivative of $y(x)$. (a) displays the original function (gray), the windowed function (black) and the sample points. The dashed line defines the window function. In (b) the analytic (gray line) and estimated numeric (points) derivative $dy(t)/dx$ is shown. The crosses indicate the result of the calculation without utilizing the window function. Note, that some crosses are vertically outside the window for $x < -2$ and $x > 2$.

With a look into line 7 of the above code, notice that the additional parameter center = T is provided to the spec.fft() function. The effect is a shift of the positive frequency spectrum to a zero-symmetric spectrum with the corresponding frequencies ranging from $-f_s/2 \leq f < f_s/2$, which enables the direct computation of $i\omega \cdot Y(\omega)$ utilizing all elements of the list variable Y. Remember, a pure DFT would map to frequencies in the range of $0 \leq f < f_s$. Here the positive frequencies $f_s/2 \leq f < f_s$ were remapped into the negative frequency domain $-f_s/2 \leq f < 0$. And in case of center = F, the purely positive frequency range would be generated and used in further calculations, which must be considered by additional steps to prevent wrong results.

2.4.2. Center a Discrete Spectrum

As initially explained in Section 2.2.3 the continuous Fourier spectrum of a continuous function $s(t)$ is defined in the range of $-\infty < \omega < \infty$. In contrast to that, its counterpart, the discrete Fourier transform, produces a 2π-periodic spectrum, which is defined for a positive number of samples $0 \leq n < N$ and $0 \leq m$ frequencies. The negative frequency components are projected into the the range of $f_s/2 \leq f < f_s$, so the spectrum becomes mirror-symmetric with respect to the Nyquist frequency $f_s/2$.

Fig. 2.8a shows the result of the continuous Fourier transform applied to the artificial example function (2.62)

$$y(t) = \sin(2\pi \cdot 4t) + 0.5 \cdot \cos(2\pi \cdot 2t) + 1.5$$

It is clear that the mean value $\overline{y(t)}$ is located at $f = 0$ and the sinusoidal components are located symmetrically around zero. In contrast to that, the discrete Fourier transform produces a data vector $\hat{Y}_m$, to which only positive discrete frequencies can be assigned. To convert the data vector into a zero-symmetric representation, the function $y(t)$ must be modulated by $f_s/2$

$$y^*(t) = y(t) \cdot e^{i\pi f_s \cdot t} \tag{2.70}$$

This method utilizes the shifting or damping property of the Fourier transform as stated in Section 2.2.3 Eq. (2.28), assuming a damping factor of $a = 1$. Doing so, the complete spectrum (Fig. 2.8b), including the mean value, shifts right towards half of the sampling frequency in the middle as indicated in Fig. 2.8a.

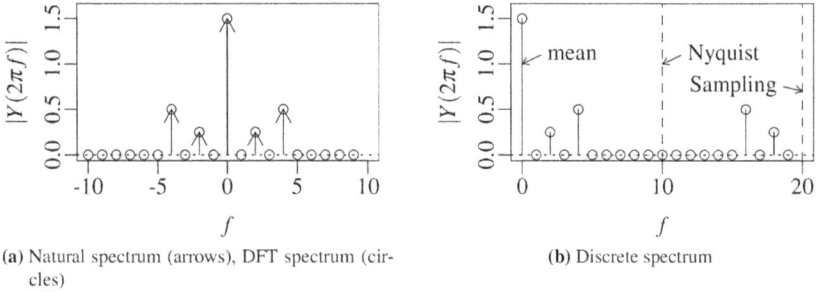

(a) Natural spectrum (arrows), DFT spectrum (circles)

(b) Discrete spectrum

Fig. 2.8. Two spectra of function (2.62). The difference between the natural spectrum in (a) and the normal discrete Fourier transform in (b) is the symmetry. In (a) there is a symmetric spectrum around zero, whereas in (b) the point of symmetry is $f_s / 2$. The dots in panel (a) indicate the result of the DFT of the modulated sampling series of (2.62). Note the exceptional first bin in (b), where the mean value is stored.

Application. In an equally spaced sampling series the discrete time vector $t = n \cdot T_s$ changes Eq. (2.70) to

$$y^*(n \cdot T_s) = \hat{y}_n \cdot e^{i\pi f_s \cdot n \cdot T_s}$$

$$= \hat{y}_n \cdot \underbrace{e^{i\pi n}}_{1-1+1-\ldots}$$

$$= \hat{y}_n \cdot (-1)^n \qquad (2.71)$$

and simplifies it. From Eq. (2.71) follows that the modulation of the data vector can be computed easily by a multiplication of the alternating series $(-1)^n$. Subsequently, all this can be extended into n dimensions, so even 2D and finally nD spectra can be centered [16, p. 154].

The following *R*-code, which is part of the `spec.fft()` function, explains the method. First, assume a vector x for the spatial location where the measurements y were taken.

```
nx <- length(x)
ny <- length(y) # is equal nx
# calculate the sampling
Ts <- min(diff(x))
# modulate the data vector
y <- y*(-1)^(1:ny)
# determine the corresponding
# frequency vector
fx <- seq(-(nx)/(2*Ts * nx)
            ,(nx - 2)/(2*Ts * nx)
            ,length.out = nx)
# calculate and normalize the spectrum
A <- fft(y)*1/ny
```

In the calculation of the frequency vector `fx` the maximum frequency is `(nx - 2) / (2*Ts * nx)` in which `(nx - 2)` considers the position of $f = 0$ and the intrinsic periodicity of the signal y. Remember the examples in Section 2.4.1, when the last possible sampling point matches the first point of the next signal period.

The back transform of the centered spectrum is the simple reverse of the described procedure. The function `spec.fft()` supports both methods to decompose a data vector into its centered or non-centered spectral representation. The usage of the code is explained below.

```
# defining the sampling
Ts <- 0.05
 x <- seq(0,1,by = Ts)
# remove last sample to avoid

# error with periodicity
 x <- x[-length(x)]
 y <- function(x) sin(2*pi * 4*x) +
       0.5 * cos(2*pi * 2*x) + 1.5

# the normal fft()
 Y <- spec.fft(y(x), x, center = F)
# prior modulation
Yc <- spec.fft(y(x), x, center = T)
# spec.fft() returns a list object
# containing the following:
# Y$fx, (Y$fy), Y$A, Y$x, Y$y, (Y$z)
```

2.4.3. The Analytic Signal

The concept of the analytic signal is based on the idea that a real-valued signal – e. g. some measured data – can only contain positive frequencies. However, at least in the 1D case, the interpretation of negative frequencies is not impossible. In addition to that, a real-world signal does only exist for positive times, usually. This makes the signal causal, which also must be taken into account, see Eq. (2.56) in Section 2.3.3. The foundation of the analytic signal is the Hilbert transform. For a short introduction on the Hilbert transform itself and its properties, refer to the section. Besides that, the attribute "analytic" originates from the Cauchy-Riemann conditions for differentiability, which is fulfilled for analytic functions in this sense [5].

Now, let's introduce the analytic signal

$$a(t) = s(t) + \mathrm{i}\mathsf{H}\left(s(t)\right) \tag{2.72}$$

It is formally defined by the sum of a signal $s(t)$ and its Hilbert transform $\mathrm{i}\mathsf{H}\left(s(t)\right)$ multiplied by the imaginary unit. Thereby the real-valued signal $s(t)$ is transformed to $a(t)$, which now has an one-sided spectrum. Frequencies above the Nyquist-frequency $f_\mathrm{s}/2$ are set to zero and frequencies below that are gained by the factor of two, so that energy is conserved. This can be seen in Fig. 2.9, which shows the result of a DFT applied on the analytic representation of the sampled signal from Fig. 2.4. Starting from that, several applications can be derived.

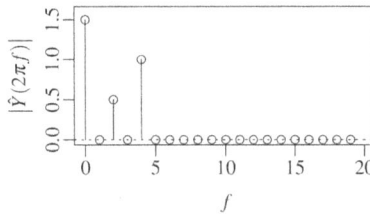

Fig. 2.9. Single sided spectrum of the analytic signal of $y(t)$.

First, all the calculations which utilize the signal's frequencies, e. g. the estimation of the derivative, introduced in Section 2.4.1, get simplified with respect to discrete data because the distinction of frequencies above and below $f_s/2$ can be omitted. The example of the derivative in

Section 2.4.1 utilized the discrete *centered* spectrum instead, which provides a frequency vector with positive and negative parts, which can be multiplied directly to the spectrum.

Second, the imaginary part of $a(t)$ equals the Hilbert transform of its real part. This enables the calculation of the envelope function because of the Hilbert transform's phase shifting properties. An instructive example will be given in the next Section 2.4.4. Besides that, the analytic signal provides a way to estimate the instantaneous frequency of a signal as shown in the example below. In consequence of that the empirical mode decomposition might be derived, which gives a time depending spectral decomposition according to Huang et al. [11]. In Section 2.4.7 it will be shown how to utilize the Fourier transform on data from non-stationary processes and how to overcome some of the FT's drawbacks in conjunction with the analytic signal.

Implementation. According to Section 2Section 2.3.3 the function analyticFunction(y) provided by the *spectral* package calculates the analytic signal. An example is given in Fig. 2.9, which displays the analytic representation of the sampled artificial example function (2.62) from Section 2.4.1. Here the spectrum is single sided and all the amplitudes have their correct value. Note that all components above $f_s / 2$ are zero, because the upper half of the spectrum is projected into the lower half by this method. In consequence, the illustration in Fig. 2.10 is in contrast to the given examples (e.g. Fig. 2.9) above, where individual amplitudes only contain the half of the true value.

The following program code explains how the analytic signal is calculated.

```
# normalized DFT-spectrum
X <- fft(x) / length(x)
# synthetic spatial vector
f <- 0:(length(X) - 1)
# shifted by half the length
# so 0 is in the middle
f <- f - mean(f)
# Hilbert transform
X <- X * (1 - sign( f - 0.5 ))
# correct mean value
X[1] <- 0.5 * X[1]

ht <- fft(X, inverse = T)
```

Here the part (1 - sign(f - 0.5)) solves three things. First, this statement shifts the phase with respect to the sign of the virtual frequency vector f, whereby the "negative" frequencies are located in the upper half of the data set. This provides a better solution than Eq. (2.36) because the evenness of the data set's length does not matter anymore. Second, since f is an integer vector, the subtraction f - 0.5 circumvents the problem $sign(0) = 0$, which avoids an error with even length data sets at $f = 0$. And third, the above code calculates the Hilbert transform and the analytic signal in one step in the spectral domain by combining Eqs. (2.35) and (2.70).

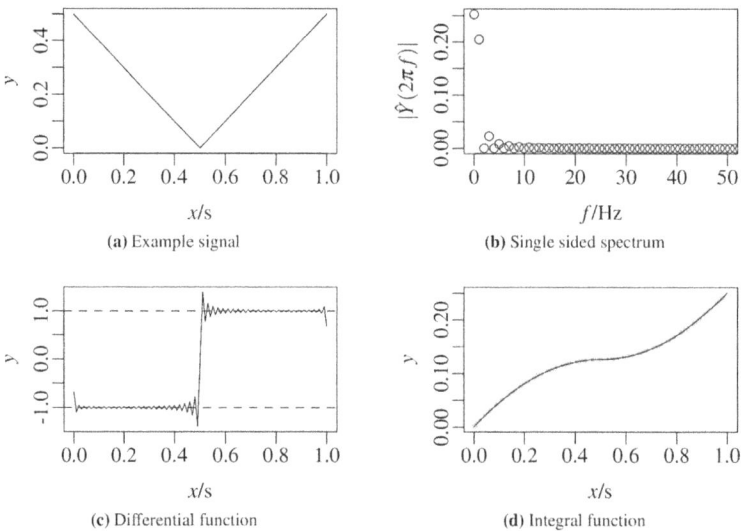

(a) Example signal
(b) Single sided spectrum
(c) Differential function
(d) Integral function

Fig. 2.10. Integral and differential calculus with an analytic signal. Dashed lines in panels (c, d) indicate the ideal analytic solution.

The code below produces the single-sided spectrum Y given in Fig. 2.9. Note, that the amplitudes correspond to the real input values of the function (2.62) and all spectral parts above $f_s / 2$ are zero.

```
x <- seq(0, 1, by = 0.05)
x <- x[ -length(x) ]

y <- function(x) sin(2*pi * 4*x) +
                 0.5 * cos(2*pi * 2*x) + 1.5
Y <- spec.fft( analyticFunction(y(x)), x, center = F )
```

Example – Integral and differential calculus. As seen in the example of Section 2.4.1, utilizing the Fourier transform on discrete data also enables fundamental calculus such as taking derivatives and integrals. In the extension of that, the analytic signal simplifies the program code which makes the whole procedure less error prone, because the symmetry of the spectrum is not important anymore.

For example, the arbitrary function $y(x)$ with

$$y(x) = |x - 0.5| \tag{2.73}$$

$$\frac{dy}{dx} = \begin{cases} -1 & x < 0.5 \\ 1 & x > 0.5 \end{cases} \tag{2.74}$$

$$\text{and } \int y(x)dx = \frac{1}{2}(x - 0.5)^2 \cdot \text{sign}(x - 0.5) \tag{2.75}$$

is steady, but not continuously differentiable. The function itself is assumed to be periodic in respect to the presented interval from Fig. 2.10.

Now, the analytic signal

$$y_{\text{ana}} = y + i\text{H}\ (y),$$

which provides a single-sided spectrum, can be used to calculate the derivative

$$\frac{dy(x)}{dx} = \text{F}^{-1}(i\omega \cdot Y)$$

by multiplying the frequency vector $i\omega$ to the corresponding Fourier spectrum $Y = \text{F}\ (y_{\text{ana}})$. The following R code illustrates the implementation, which uses the uncentered spectrum because the analytic signal is already single-sided.

```
# calculate the non-centered spectrum
Y <- spec.fft(x = x, y = y, center = F)

# taking the derivative in frequency space
Y$A <- (1i*2*pi*Y$fx) * Y$A
# back transform
yd <- Re( spec.fft(Y)$y )
```

Fig. 2.10c depicts the result of the calculus above. It is clearly visible, that the (unsteady) step-function – which is composed of an infinite spectrum – causes typical ringing effects in the region of the jump.

In contrast to that, the result of integral calculation shown in Fig. 2.10d remains smooth. The implementation is almost straight forward like the differentiation. To retrieve the integral the analytic function is emphasized again, so that

$$\int y(x)dx = \mathsf{F}^{-1}\left(\frac{1}{i\omega} \cdot Y\right) = \mathsf{F}^{-1}\left(\frac{1}{i\omega} \cdot Y\right)\bigg|_{\omega>0} + |Y(0)| \cdot x$$

would produce the result. But this method must be tuned, since the first element of the DFT contains the mean value of of the data, which would be divided by 0 when multiplying the frequency vector. The code snippet below elucidate the procedure on how to circumvent this issue.

```
# calculate the non-centered spectrum
Y <- spec.fft(x = x, y = y, center = F)

# saving the mean
m <- Y$A[1]
# taking the derivative in frequency space
Y$A <- 1/(1i*2*pi*Y$fx) * Y$A
Y$A[1] <- 0 # correct the div/0 in the first bin

# back transform
yd <- Re( spec.fft(Y)$y ) + m * x
```

Example – Instantaneous frequency and amplitude. Another advantage of analytic signals is that the real part (the input signal) and the imaginary part (as provided by Hilbert transform) are not calculated once for a time interval, but for *every* sample. This allows the calculation of the instantaneous amplitude $A(x)$ and instantaneous frequency $f(x)$, as defined below:

$$A(x) = \sqrt{\Re\left(y_{ana}(x)\right)^2 + \Im\left(y_{ana}(x)\right)^2} \qquad (2.76)$$

$$\varphi(x) = \arctan\frac{\Im\left(y_{ana}(x)\right)}{\Re\left(y_{ana}(x)\right)} \qquad (2.77)$$

$$f(x) = \frac{1}{2\pi}\frac{d\varphi}{dx} \qquad (2.78)$$

The application of this principle is shown in the function definition

$$y(x) = \sin(\pi x)^2 \cdot \sin\left(2\pi \cdot 4 \cdot x + \frac{1}{2\pi}\cos(2\pi \cdot x)\right),$$

which includes amplitude and frequency modulation simultaneously, as seen in Fig. 2.11. By definition, the signal is band-limited to facilitate the calculations. Non-band-limited data has to be weighted with an appropriate window function to preserve conditions (2.20) to (2.22). It can be clearly seen that the frequency modulation as well as the amplitude modulation can be extracted from the discrete data. The following code fragment describes the application in *R*.

```
# define modulation parameters
fo = 4

fm = 1
am = 1/(2*pi*fm)

x <- seq(0, 1, by = 1/50)
y <- sin(pi*x)^2 * sin(2*pi*fo*x + am * cos(2*pi*fm*x))

ya <- analyticFunction(y)

yphi <- atan2(Im(ya),Re(ya))

# the phase is output in the range
# -pi ... pi and must be unfolded
dyphi <- c(0,diff(yphi))
dyphi_max <- 0.9 * max(abs(dyphi))

tmp <- yphi
n <- 0
for(i in 1:length(dyphi) )
{
   if(dyphi[i] < -dyphi_max)
   n <- n + 1
   yphi[i] <- tmp[i] + n*2*pi
}

# time dependent amplitude and frequency
f <- c(NA, diff(yphi) / min(diff(x))) / (2*pi)
A <- sqrt( Re(ya)^2 + Im(ya)^2 )
```

(a) Signal

(b) Amplitude-frequency map

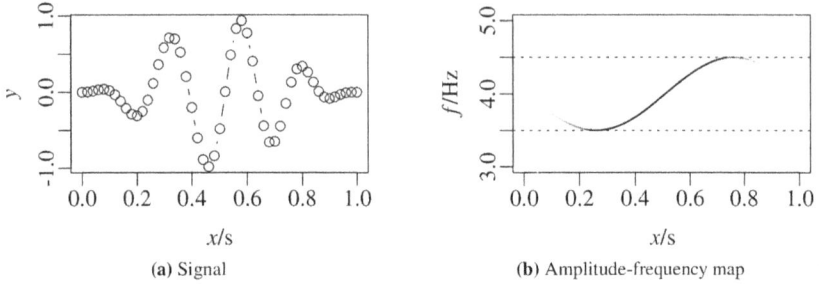

Fig. 2.11. Instantaneous frequency and amplitude. Panel (a) introduces an amplitude and frequency modulated signal. Circles indicate sampling points. The amplitude intensity is grey scale-coded in panel (b).

The example above illustrates the procedure of determining the instantaneous amplitude and frequency of a single mode signal, which is defined as a signal with only one carrier signal, varying in $A(x)$ and/or $f(x)$. A generalization of this concept for arbitrary (multimode) signals is presented in the work of Huang et al. [9] with the empirical mode decomposition.

Because of the special nature of the Hilbert transform the frequency modulation can be resolved quite precisely. The leakage effect, which occurs in ordinary DFT or shifting DFT, appears much less pronounced or can even be circumvented completely, as can be seen in the example above. Here the maximum DFT frequency resolution is $\delta f = 1$, if *all* the data set is used, which makes it impossible to track a 10 ± 0.5 Hz wobbling signal. However, the investigation by means of analytic signals enables such a detailed analysis.

2.4.4. Calculating the Envelope

Sometimes it becomes necessary to calculate the envelope function of data. Different approaches are possible, for instance finding all maxima and then fitting a spline function to these points. However, in the following the spectral way is introduced.

First, remember the trigonometric identity

$$\sin^2(x) + \cos^2(x) = 1, \tag{2.79}$$

which becomes the key component of the calculation later. Next to that, assume a function

$$y(t) = A(t) \cdot \cos(\omega_0 t) \qquad (2.80)$$

with $A(t)$ as the envelope function, which is modulated with the carrier $\cos(\omega_0 t)$. It is clear now that calculating the envelope is equivalent to an amplitude demodulation.

Now, the amplitude function is gained by

$$A(t) = \sqrt{y(t)^2 + y^*(t)^2}$$

$$= A(t) \cdot \underbrace{\sqrt{\cos^2(\omega_0) + \sin^2(\omega_0)}}_{=1},$$

where $y^*(t)$ denotes the signal $y(t)$ phase shifted by $\pi/2$, so that $y^*(t)$ equals the Hilbert transform

$$y^*(t) = A(t) \cdot \sin(\omega_0 t) \qquad (2.81)$$

$$= \mathsf{H}\left(y(t)\right) \qquad (2.82)$$

Under the condition of *sufficient* band limitation the statement above can be expressed as

$$A(t) = \sqrt{y(t)^2 + \mathcal{H}\left(y(t)\right)^2}, \qquad (2.83)$$

which works, because (2.82) functions as an ideal phase-shifter as discussed in Sections 2.2.3 and 2.3.3.

Application. The calculation is straightforward and follows Eq. (2.83). The *spectral* package includes the function envelope() to perform the calculation of an 1D envelope. The only problem with this is band limitation, which is necessary to achieve reasonable results. Compare Fig. 2.12b, here the required bandwidth to demodulate the envelope becomes clearly visible in the negative and positive half plane.

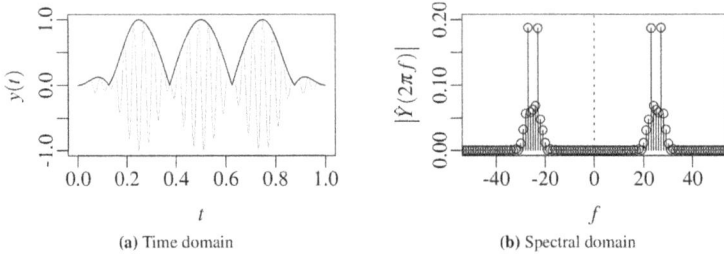

(a) Time domain   (b) Spectral domain

Fig. 2.12. The envelope function of a signal. The spectrum (b) clearly indicates the band width, which is needed to reconstruct the envelope. In case of a noisy signal the band-pass filter *must* provide at least this width to obtain reasonable results.

2.4.5. Convolution

The Fourier transform does not only map a spatial function into the spectral domain, moreover it also converts several mathematical operations. One example is the derivation of

$$df(t) / \ dt \leftrightarrow i\omega F(\omega), \tag{2.84}$$

which equals a simple multiplication with the complex variable $i\omega$ in the spectral domain as seen in the example of SubSection 2.4.1.

In the following, the convolution $\int_{-\infty}^{\infty} f(\tau)g(t-\tau) \ d\tau$ will be introduced by the identities

$$F\left(\underbrace{\int_{-\infty}^{\infty} f(\tau)g(t-\tau)d\tau}_{=f(t)*g(t)}\right) = \tag{2.85}$$

$$= F\ (f(t))\cdot F\ (g(t)) \text{ in the spatial domain}$$

$$\text{and } F\ (f(t)\cdot g(t)) =$$

$$= \int_{-\infty}^{\infty} F(\tau)G(\omega-\tau)d\tau \text{ in the spectral domain,} \tag{2.86}$$

which are valid only if the integrals $\int_{-\infty}^{\infty}|f(t)|^2 \ dt < \infty$ and $\int_{-\infty}^{\infty}|g(t)|^2 \ dt < \infty$ exist. Eq. (2.85) describes the convolution operation in the spatial domain, which corresponds to a simple multiplication in spectral domain.

The working principle and the implementation of (2.85) and (2.86) via the fast Fourier transform (FFT) do form one of the most powerful tools in the field of numeric computation.

Example – Polynomial multiplication. Suppose two polynomials of degree $N = 2$. Then the multiplication of these two would be

$$\left(a_0 + a_1 x + a_2 x^2\right) \cdot \left(b_0 + b_1 x + b_2 x^2\right) =$$

$$= a_0 b_0 x^2 a_0 b_0 + a_0 b_0 x^2 a_0 b_1 x + a_0 b_0 x^2 a_0 b_2 x^2 +$$

$$+ a_0 b_0 x^2 a_1 b_0 x + a_0 b_0 x^2 a_1 b_1 x^2 + a_0 b_0 x^2 a_1 b_2 x^3 +$$

$$+ a_0 b_0 x^2 a_2 b_0 x^2 + a_0 b_0 x^2 a_2 b_1 x^3 + a_0 b_0 x^2 a_2 b_2 x^4 + = \sum_{i=0}^{2N} c_i x^i, \quad (2.87)$$

which finally gives an expression of degree of $2N = 4$. Performing the expansion will end up in a convolution of the two coefficient vectors, which must be $2N + 1$ elements long to fit the result in the output vector. Doing all that will require $(N + 1)^2 = 9$ floating point multiplications on a processor. The statement

$$\begin{pmatrix} a_0 \\ a_1 \\ a_2 \\ 0 \\ 0 \end{pmatrix} * \begin{pmatrix} b_0 \\ b_1 \\ b_2 \\ 0 \\ 0 \end{pmatrix} = \begin{pmatrix} a_0 b_0 \\ a_0 b_1 + a_1 b_0 \\ a_0 b_2 + a_1 b_1 + a_2 b_0 \\ a_1 b_2 + a_2 b_1 \\ a_2 b_2 \end{pmatrix} \quad (2.88)$$

$$\underbrace{\qquad}_{\vec{a}} \quad \underbrace{\qquad}_{\vec{b}} \quad \underbrace{\qquad\qquad\qquad}_{\vec{c}}$$

expresses the convolution and its result, where the symbol * denotes the convolution operator. Alternatively, this operation can be performed with the help of the Fourier transform. According to (2.85)

$$\vec{c} = \mathsf{F}^{-1}\left(\mathsf{F}\left(\vec{a}\right) \cdot \mathsf{F}\left(\vec{b}\right)\right)$$

requires the transform to be calculated three times.

The great advantage of using the Fourier transform, instead of the straightforward expansion, is the scaling of $N \cdot \log_2 N$ for the implementation of the FFT [17, Chapter 30] in contrast to the N^2 scaling of conventional factor expansion. As shown in Table 2.2 the FFT accelerates the computation from a value of $N > 32$ for this example.

Table 2.2. Computational costs for the polynomial multiplication.
The calculation amount for the FFT is estimated by $3 \cdot (2n+1)\log_2(2n+1)$.

Degree	Expansion	FFT
8	64	209
16	256	500
32	1024	1175
64	4096	2714

The standard algorithm of the discrete FFT is limited by a vector size of $N = 2^n$, which can be overcome with the implementation of the DFT method as described by M. Frigo [18], which even allows prime numbered lengths without any loss of performance.

As a simple example the following code illustrates the method described above.

```
# defining the coefficients
a <- c(1:3, 0, 0)
b <- c(5:7, 0, 0)
# calculating the FFT
A <- fft(a)
B <- fft(b)

# convolve via multiplication
# of A and B
c <- Re( fft(A * B, inverse=T)
/ length(a) )
# would give the same result
c2 <- convolve(a, b, conj = F)
print(c)
####### OUTPUT ########
> 5 16 34 32 21
```

The result is given in the last line of the program listing. The convolve(x, y, conj = F) *R*-command uses the same mechanism and produces exactly the same results.

2.4.6. The Sampling Theorem and Filtering

Every sampled signal has its starting point $t = 0$, followed by a point in time T where it ends. The sampling procedure itself and the nature of the signal finally define the underlying band limitation. The intention of the first part of this section is to show the relation between the measurement and the consequences which arise if a causal signal is sampled. It turns out that in conjunction with the Fourier transform several convolutions take place in the time and frequency domain by selecting the maximum time T and the sampling frequency.

Example – The sampling process. Regarding that, it is important to understand that the filtering process already begins with the sampling of the signal. Fig. 2.13 illustrates that for the function

$$y(t) = 0.8 \cdot \cos(2\pi \cdot 2t), \tag{2.89}$$

which fits perfectly into the period of the sampling window. In principle, this function is defined on the interval of $-\infty < t < \infty$. Then, the data acquisition process defines a starting point at $t = 0$ where the measurement begins and an end point $t = T$ where the measurement stops. This is illustrated by the bold rectangular window

$$w(t) = \begin{cases} 1, & 0 \le t < T = 1 \\ 0, & \text{elsewhere} \end{cases} \tag{2.90}$$

in Fig. 2.13a. This window function picks a range $0 \le t < T$ of $y(t)$ so that the signal

$$s(t) = w(t) \cdot y(t) \tag{2.91}$$

is now defined by $y(t)$ multiplied by $w(t)$. According to the convolution properties of the Fourier transform (see Section 2.4.5 and Eq. (2.31)), this operation corresponds to the convolution

$$S(\omega) = \int_{-\infty}^{\infty} W(\omega - \tau) \cdot Y(\omega) d\tau = W(\omega) * Y(\omega) \tag{2.92}$$

of the ideal spectrum $Y(\omega) = \mathcal{F}(y(t))$ (which are two Dirac impulses) with the spectral representation

$$W(\omega) = \frac{1}{i\omega} \left(1 - e^{-i\omega T} \right) \tag{2.93}$$

of the window function $w(t)$, which is in fact Shannon's sampling function [13]. The consequence for the signal's spectrum $S(\omega)$ is, that the ideal Dirac-spectrum $Y(\omega)$ smears out to $\sin(x)/x$-like functions, because each arrow is weighted with $W(\omega)$. Note, the zeros of the resulting spectrum are in an equidistant spacing of $\delta f = 1/T$ now. At this moment the original function $y(t)$ is just windowed and not yet sampled.

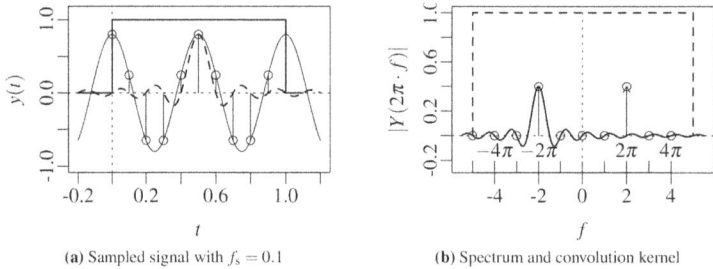

(a) Sampled signal with $f_s = 0.1$ (b) Spectrum and convolution kernel

Fig. 2.13. In (a) the function $y(t)$ from Eq. (2.89) (thin line) is sampled in the time range $0 \leq t < 1$. The finite time interval corresponds to a rectangular weight function (bold line). The continuous spectrum $Y(\omega)$ (arrows in (b)) is then convoluted with the $\sin(x)/x$-like spectral representation of that rectangular window. The bold line indicates the convolution step for the left spectral part. Note that only the Dirac impulse generates a value, because all other components remain zero. The dashed line in (b) indicates the limitation to a maximum frequency, which corresponds to a convolution in the time domain (a). Open symbols show the sampling (a) and the result of the DFT (b).

In the next step the spectrum in Fig. 2.13b is limited to a maximum frequency of $f_{max} = 1/2T_s$ by applying an additional rectangular window in the frequency domain

$$V(\omega) = \begin{cases} 1, & |\omega| \leq 2\pi \cdot (2T_s)^{-1} \\ 0, & \text{elsewhere} \end{cases} \qquad (2.94)$$

The parameter T_s represents the period of the sampling frequency. The important link to the Nyquist frequency will be clarified later. However, the corresponding representation of $V(\omega)$ in the time domain

$$v(t) = \frac{\sin\left((2T_s)^{-1} \cdot t\right)}{t} \qquad (2.95)$$

is displayed by the dashed line in Fig. 2.13a. In the context of the convolution, which now takes place in time domain, the windowed signal

$$s(t) = \left(w(t) \cdot y(t) \right) * v(t) \qquad (2.96)$$

is finally convoluted with $v(t)$. But still, the result (2.96) is not sampled yet, which is carried out in the next step.

Sampling is a procedure where individual values of the function $s(t)$ are selected and stored. With regard to the nature of the sinc-like functions in Fig. 2.13, the sampling should take place at each possible zero of $v(t)$, as well as $v(t = 0)$ which is the only sampling point with a non-zero value of $v(t)$. This behavior guarantees that the convolution (2.96) rejects everything but the points of sampling. In the same moment also the distance in between them is set to $\delta t = T_s$, because the function $v(t)$ provides two zeros in an interval of $(2T_s)^{-1} \cdot t \leq 2\pi$, which correspond to the next non-zero samples of the neighboring points. In other words, by choosing the next sample $v(t)$ must be shifted by exactly T_s to reject the former sample and use the current value. Note, violating this restriction, because the location of sampling points jitters, will lead to an error in the discrete spectral representation $\hat{S}_m$.

Finally, this conceptual description gives a second access to the question why the sampling frequency should be at least twice the maximum signal frequency. A detailed mathematical derivation of the discrete Fourier transform and the properties of sampling is given in the book of L. Debnath [19, Chapter 3]. However, the discrete sampling given in Fig. 2.13a and the finite length of the sampling series causes the convolution (2.92) to take the only possible peaks in the spectrum at $f = \{-2, 2\}$, because all other 2π-periodic values evaluate exactly to zero.

If the signal period does not fit perfectly into the window, which is mostly the case in reality, the kernel function $W(\omega)$ never matches the true Dirac, but instead of that it sums up values into the left and right bin of the corresponding discrete Fourier vector $\hat{S}_m$. To reduce these side effects, other window functions can be used in advance, before the DFT takes place. The intent of the window functions (Hamming-, Blackman- or Tukey-window) is to convolve the spectrum with a steeper decreasing function amplitude compared to $W(\omega)$ to minimize the effect of the period's misfit. Moreover $w(t)$ has an infinite spectrum, which is then truncated in the finite and discrete data processing, which leads to additional side effects if the period does not fit the sampling window exactly. However, window functions can help to reduce such interference, which are generated when local non-periodic events or infinite spectra are present.

Example – convolution filter. Convolution filter in the time domain, such as the moving average filter, are easy to calculate with the help of the FFT instead of convoluting a filter kernel along the signal. Given the signal

$$s(t) = \cos(4\pi \cdot t) + \sin(20\pi \cdot t) + \mathcal{N}(0, 0.5)\big|_{0 \leq t < 1},$$

which is defined in the range of $0 \leq t < 1$. The moving average of the sampled signal $\hat{s}_n$ with the filter kernel length of $N_K = 5$

$$\overline{\hat{s}_n} = \sum_{i=1}^{N_K} \hat{k}_i \cdot \hat{s}_{n-i} \tag{2.97}$$

is then calculated for the nth element by taking the weighted sum of N_K elements before that element.

The filter kernel $\hat{k}$ for the example in Fig. 2.14 is represented by the weighting coefficients

$$\hat{k} = \frac{1}{5} \begin{pmatrix} 1 \\ 1 \\ 1 \\ 1 \\ 1 \end{pmatrix} \tag{2.98}$$

With respect to the spectrum of $\hat{k}$, it becomes clear that the length of $\hat{k}$ is chosen in a way that the high frequency part $\sin(20\pi \cdot t)$ is rejected completely. But it is also evident that other frequency components still remain in the result or were damped unintentionally, like the low frequency term $\cos(4\pi \cdot t)$.

Fig. 2.14a illustrates that the application of the moving average filter in the time domain will also lead to a phase-shift in the result. Working in the spectral domain, one will gain the possibility to shift the signal back and reconstruct the whole definition range as follows by utilizing the FT shifting property:

$$\mathsf{F}\left(\overline{s(t)}\right) = e^{i\omega\frac{N_K}{2}} \cdot S(\omega) \cdot K(\omega) \tag{2.99}$$

$$\mathsf{F}\left(\overline{\hat{s}(n)}\right) = e^{i2\pi\frac{(0..(N-1))\cdot N_K}{2N}} \cdot \hat{S}_m \cdot \hat{K}_m \tag{2.100}$$

The result of this operation illustrates the dash-dotted line in Fig. 2.14a. The code for this calculation is presented below.

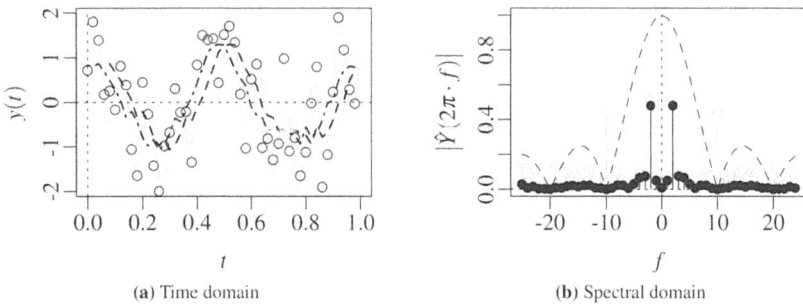

(a) Time domain   (b) Spectral domain

Fig. 2.14. Calculating the moving average to reduce the noise and to reject the high frequency. Panel (a) shows the original function (gray line) and the noisy sampling points (open symbols). The result of the moving average in the time domain is given in (a) for the convolution (dashed) and the shifted FFT (dash-dotted). (b) displays the spectrum of the original function (gray), the filtered result (black) and the kernel function (dashed).

```
# Convert the kernel k into
# its spectral representation.
# The different lengths are adapted.
K <- fft( c(k,rep(0, length(y)-length(k))))

exp_ex <- exp( 2i*pi * (1:length(y)-1) /
length(y) * length(k) %/%2 )
Y <- fft( y ) / length(y)
y_spec <- Re( fft( exp_ex * Y * K, inverse=T) )
```

Pay attention to the statement `fft(c(k, rep(0, length(y) - length(k))))` in the first line. It first fills K with zeros up to the length of the whole data set, before the FFT is calculated. This is because the multiplication in frequency space can only take place between vectors of equal lengths.

With respect to the spectrum of the kernel $\hat{K}$ – as seen in Fig. 2.14b – it becomes evident that the moving average filter is not able to suppress higher frequencies completely.

Example – The ideal low pass filter. Since the moving average has some drawbacks (because of its infinite frequency response) one should think about an ideal (low pass) filter with a finite frequency response. Such a filter has the advantage that unwanted spectral components can be rejected completely. On the other hand, this results in an infinite long kernel in the time domain, which is nevertheless calculated in the spectral domain.

The basic principle of such a filter is to set each component above a certain frequency $|\omega_{max}|$ value to zero

$$S(|\omega| > \omega_{max}) = 0 \tag{2.101}$$

Fig. 2.15 illustrates the procedure. Compared to the previous moving average example in Fig. 2.14, the output signal's shape changes to a more smooth and accurate form. However, there remains a difference compared to the ideal low frequency part (dashed line in Fig. 2.15a), which is the result of the remnant spectral components around the signal frequency $f = 2$.

A code snippet to calculate the ideal low pass is presented below.

```
# assuming x and y to hold the time
# vector and the sampling points

Y <- spec.fft(y, x)
# doing the filtering
Y$A[abs(Y$fx) > 3] <- 0

# reconstruct the filtered signal
yf <- spec.fft(Y, inverse = T)$y
```

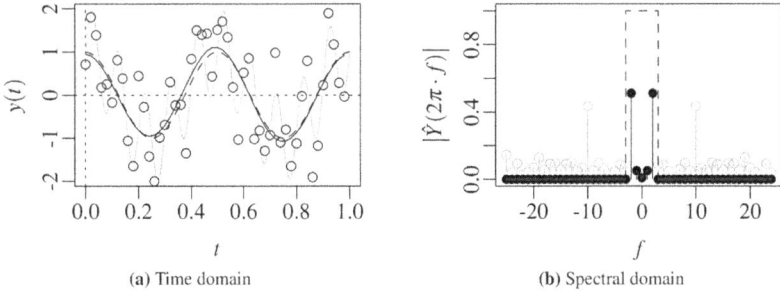

(a) Time domain   (b) Spectral domain

Fig. 2.15. Calculating the ideal low pass filter. Panel (a) shows the original function (gray line) and the noisy sampling points (open symbols). The result of the ideal low pass filter in the time domain is given in (a) by the solid line. The dashed line in (a) illustrates the low frequency part of the signal with $f = 2$. Panel (b) displays the spectrum of the original function (gray), the filtered result (black) and the spectral kernel function (dashed).

The *spectral* package provides the `filter.fft(y, x, BW, fc = 0, n = 3)` function to do that in one step. It utilizes the `analyticFunction(y)` to filter the one-sided spectrum and enables the user to apply a band-pass filter with the arbitrary polynomial weight function

$$w \propto \omega^n \qquad\qquad (2.102)$$

The parameter n sets the polynomial degree, whereas 3 is close to the moving average. Passing higher values, say `n = 10`, will successively converge to the ideal band-pass solution. This enables the user to choose between a smooth frequency roll-off or a steep decline, depending on the application.

The used code then changes to the following example

```
yf <- filter.fft(y, x, fc = 0, BP = 3, n = 10)
```

As seen in Fig. 2.15 the bandwidth BP is symmetric around the center frequency `fc = 0`, so the values must be set in accordance to the desired window which should remain in the spectrum. Utilizing the analytic signal representation avoids mistakes when programming the weight function and keeps the code clean.

101

Example – Noise reduction via autocorrelation. By choosing an appropriate kernel function it becomes possible to extract certain features of the signal. The previous example illustrated how a low pass filter can suppress the noisy components of the signal by zeroing all upper frequency components in the spectrum. In the time domain this corresponds to a convolution operation of the input data with an infinite sinc-like kernel function.

Now the opposite approach is to ask what the most significant periodic components within the signal are. First of all, the continuous autocorrelation function

$$acf(\tau) = \int_{-\infty}^{\infty} s(t)s^*(t-\tau)\,\mathrm{d}\tau \qquad (2.103)$$

$$= \mathsf{F}^{-1}\left(S(\omega)\cdot S^*(\omega)\right) \qquad (2.104)$$

provides a mechanism to examine a data set or function with respect to its self-similarity. The equations above convolute the signal $s(t)$ with its complex conjugate $s^*(t)$. According to the convolution properties, Eq. (2.102) can be expressed as the inverse Fourier transform of the product of the signals spectra. Note, for a real-valued signal $s(t) = s^*(t)$. Suppose the signal consists of a stationary sinusoidal signal. It is quite evident that acf rejects the random noise which overlays $s(t)$. The underlying noise equals a non-stationary process and is therefore canceled out by utilizing the acf. A generalization of this statement is the Wiener-Khinchin theorem. Its application is discussed in detail by L. Cohen [5, Chapter 1.9], R. Marks [20, Chapter 4] and N. Wiener [21].

Assuming the function

$$s(t) = \cos(2\pi \cdot 2 \cdot t) + \sin(2\pi \cdot 10 \cdot t) + \mathcal{N}(0, 0.5) \qquad (2.105)$$

with some superimposed normal-distributed noise. The signal $\hat{s}_n$ is then the sampled function $s(n \cdot T_s)$ with $T_s = 0.01$, so it fits perfectly into one sampling period of $T = 1$. This example is again very artificial. Real world signals often cause aliasing effects, so attention must be paid.

The discrete spectrum of the acf_n can now be calculated by

$$ACF_m = \hat{S}_m \cdot \hat{S}_m^* \qquad (2.106)$$

Note, the resulting spectrum is real-valued, so the phase information will be lost. A proper noise reduction can be achieved by defining a weighting vector

$$
W_m = \begin{cases} 0, & |ACF| < \mathrm{sd}(|ACF_m|) \\ 1, & \text{else} \end{cases}, \tag{2.107}
$$

which sets every spectral component smaller than the standard deviation sd of the ACF_m to zero. The resulting filtered signal

$$
\hat{s}_{\mathrm{f}}(n) = \mathsf{F}^{-1}\left(W_m \cdot \hat{S}_m\right) \tag{2.108}
$$

is given in Fig. 2.16.

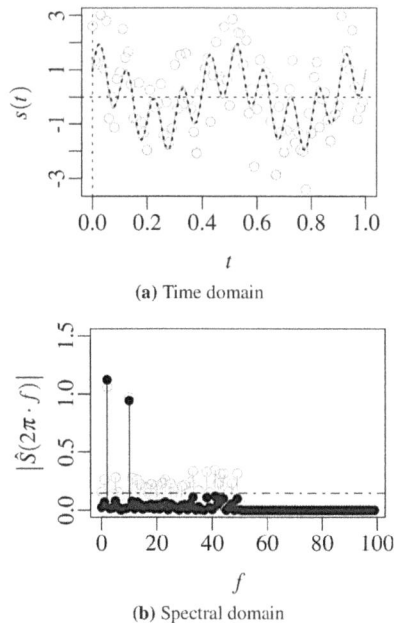

(a) Time domain

(b) Spectral domain

Fig. 2.16. Noise reduction with autocorrelation. In the left panel (a) the sampled function (gray line) is overlaid with noise (circles). The bold line is the result of the weighted filtering procedure. In the frequency domain (b) the single sided spectra of the analytic functions are displayed. The input function (gray symbols) contains a lot of noise which is reduced due to autocorrelation (dotted black). The horizontal line indicates the threshold below which all spectral components are ignored.

In the example above, it is important to calculate everything with the analytic function representation to preserve the energy content of the data vectors.

```
x <- seq(0, 1, by = Ts)
# for perfect periodicity
x <- x[ -length(x) ]
y <- cos(2*pi * 2*x) + sin(2*pi * 10*x)
      + rnorm(length(x), sd = 1)

# calculating the autocorrelation
Y <- fft( analyticFunction(y) ) / length(y)
ACF <- Y * Conj(Y)

# calculating the weight vector
w <- ACF
w[abs(w) < sd( abs(w) )] <- 0
w[w != 0] <- 1
# backtransform of the filtered signal
yf <- Re(fft( Y * w, inverse = T) )
```

2.4.7. Non-Stationary Processes – Spatially Dependent Spectral Analysis

Many signals in reality do not fulfill the requirement of stationarity. It is very often the case that the measured signal is overlaid by a trend or a temporal-local event that takes place only once. All these in-stationary and non-periodic processes will spread out into all frequencies in the corresponding Fourier decomposition of the data. If proper band limitation cannot be achieved, the mapping between physical frequencies and the spectral representation might fail too, as stated in Section 2.3.2. Remember that the Hilbert transform and the associated concept of instantaneous frequencies and amplitudes could also be applied here as demonstrated in Section 2.4.3.

The equation

$$y(t) = \underbrace{\frac{1}{\sqrt{2\pi \cdot 0.05}} \cdot e^{-\frac{(x-0.2)^2}{2 \cdot 0.05^2}}}_{A_1(t)} \cdot \sin(2\pi \cdot 20 \cdot t) +$$

$$+\underbrace{\frac{1}{\sqrt{2\pi \cdot 0.1}} \cdot e^{-\frac{(x-0.7)^2}{2 \cdot 0.1^2}} \cdot \sin\left(2\pi \cdot 40 \cdot t\right)}_{A_2(t)} \qquad (2.109)$$

describes an arbitrary example of a non-stationary signal, which is illustrated in Fig. 2.17a.

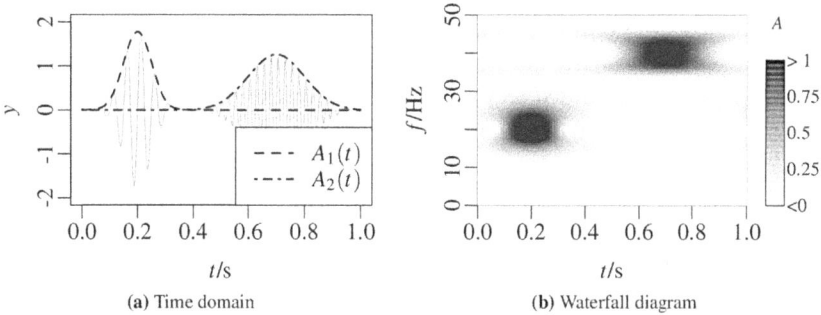

(a) Time domain (b) Waterfall diagram

Fig. 2.17. Basic simple example of a non-stationary signal (a) with its time depending spectral decomposition in (b). The different line types of the envelope correspond to the first and second event.

Here two different events occur. The first one has a low frequency at $f_1 = 20\,\text{Hz}$, whereas the second one oscillates faster with $f_2 = 40\,\text{Hz}$. Both parts are modulated with a Gaussian envelope to switch each of them on and off. Finally, the right panel (b) shows the resulting time variant decomposition of the signal $y(t)$. The waterfall() function from the *spectral* package can be used to calculate such kinds of waterfall diagrams.

Note, this type of analysis produces a two-dimensional time-frequency map like a shifting, or short-time FFT, which – in contrast to the waterfall analysis – selects only a window of the data before doing the spectral decomposition for one time step. For a better understanding of this special type of signal analysis lets assume the amplitude envelope functions $A_1(t)$ and $A_2(t)$ as arbitrary window functions which mask the time series. The interpretation is that $A_i(t)$ contains information in time, where the finite process is located. All this relies on the assumption that the signal of interest can be modeled as

$$y(t) = \sum_i A_i(t) \cdot \cos(\omega_i \cdot t + \varphi_i) \qquad (2.110)$$

In contrast to the shifting FFT, the introduced approach is to do an amplitude demodulation by calculating the signal's envelope function $A(t)$ for a given center frequency f_c. The latter corresponds then to the frequency of the physical process. Finally, a sufficient bandwidth around f_c guarantees that the shape of $A(t)$ will be reconstructed correctly.

The first advantage of this method is that the whole data set is taken into account for each single frequency of interest. Note, in contrast to that, the shifting FFT would select only a certain range of the data at once, so information about low frequencies would be lost if the selected window becomes too small. However, the overall frequency resolution decreases to the length of the window size of the shifting FFT. In the end, two signals that are very close in the frequency space might not be distinguishable anymore. The second point is the necessary band-pass filter at f_c, which is changed according to the actual frequency. This draws attention to the uncertainty principle

$$2\pi f \cdot T = 1,$$

which states that with increasing frequency the locating of an event becomes sharper but the exact frequency gets more incorrect.

Implementation. As stated above, the key component is a band-pass filter BP, which must be applied twice to calculate the two envelopes in Fig. 2.17. The following code shows how this can be done in *R* with the appropriate functions provided by the *spectral* package.

```
A1 <- Re(filter.fft(y, x, fc = 20, BW = 10, n = 10)),
A1 <- Re(envelope(A1)),
A2 <- Re(filter.fft(y, x, fc = 40, BW = 10, n = 10)),
A2 <- Re(envelope(A2)).
```

The next step is to do this calculation for many frequencies. A reasonable range is to start from $f = 0$ up to $f_s/2$. To accelerate the code the waterfall() function uses a fast version of the envelope() function, which combines the filtering and the Hilbert transform as follows.

```
# Defining the frequency vector
Y.f <- seq(0,(n-1)*df,length.out=n)
# calculating the sign for the HT
sY.f <- (1 - sign(Y.f-mean(Y.f)))
# correct first bin
sY.f[1] <- 1

fast_envelope <- function(y,x,fc,BW,nf)
{
  Y <- BP(Y.f,fc,BW,nf) * fft(y) / length(y) * sY.f
  hk <- base::Mod( fft(Y + 1i*Y,inverse=T) / sqrt(2) )
  return(hk)
}
```

Here the working band-pass filter is implemented as a weighting vector multiplied by the amplitudes. The second line in the fast_envelope() function performs the back transform. Compared to the calculation of A1 and A2 this method saves two back transforms and one real part extraction and thus accelerates the calculation.

Finally, the bandwidth calculation is done by an empirical approach, like

$$BW(f_c) = \begin{cases} 4\delta f & f_c < 16\delta f \\ f_c/4 & 16\delta f \le f_c \le wd \cdot \delta f \\ wd \cdot \delta f & \text{else} \end{cases} \qquad (2.111)$$

Here $\delta f = 1/\Delta x$ denotes the frequency step and wd is the normalized width of the resulting band-pass. The task of $BW(f_c)$ is to widen the frequency band for higher frequencies, whereas low frequencies take a very small band width. This takes the uncertainty principle into account, which states that "a narrow waveform yields a wide spectrum and a wide waveform yields a narrow spectrum and both the time waveform and frequency spectrum cannot be made arbitrarily small simultaneously" [22].

Application. The two following examples illustrate how the waterfall() function can be used for intricate signals. First of all the function

$$y(t) = |2t - 1| \cdot \sin\left(2\pi \cdot 10 \cdot t\right) + \begin{cases} 0 & t \le 0.5 \\ \sin\left(2\pi \cdot 20 \cdot t^2\right) & \text{else} \end{cases} \qquad (2.112)$$

is going to be analyzed.

Fig. 2.18 illustrates the results. The waterfall diagram in panel (c) displays all the features of Eq. (2.112). Note that even the $|x|$ function and the $\sin(x^2)$ term can be distinguished, whereas the normal time-invariant Fourier spectrum in Fig. 2.18c hides this property completely. Here the 10 Hz carrier is the only "correctly" visible signal component.

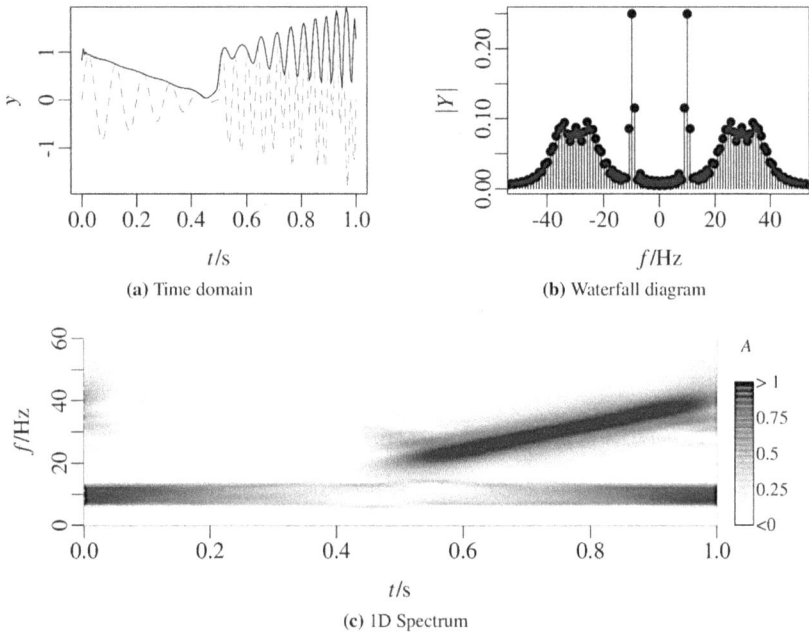

(a) Time domain

(b) Waterfall diagram

(c) 1D Spectrum

Fig. 2.18. Complex example for the temporal-dependent spectral decomposition. (a) shows the time series and (c) its decomposition. Panel (b) demonstrates that the simple 1D spectrum would produce misleading results. Here one can only identify the 10 Hz carrier, but the frequency drift is hidden behind the wide band spectrum at around 30 Hz .

Example – Temperature response of a 300 kW cooling system. A second example is given in Fig. 2.19a. Here the data represents the outflow temperature of a cooling system with about 300 kW power. It consists of a heat exchanger outside the building and a secondary water loop inside from which to draw the heat. A sprinkler system improves the performance of the outside heat exchanger, so that outflow temperatures below the ambient temperature become possible. Panel (b) illustrates the ambient temperature near the external air heat exchanger. The impact of varying solar radiation becomes visible in a signal amplitude with a very long period of about $T_{sun} \approx 80$ min. In terms of period lengths the high eigenfrequencies of the hydraulic system become visible at the bottom, approximately within a range of 10 min to 20 min, when the control loop reaches its limit.

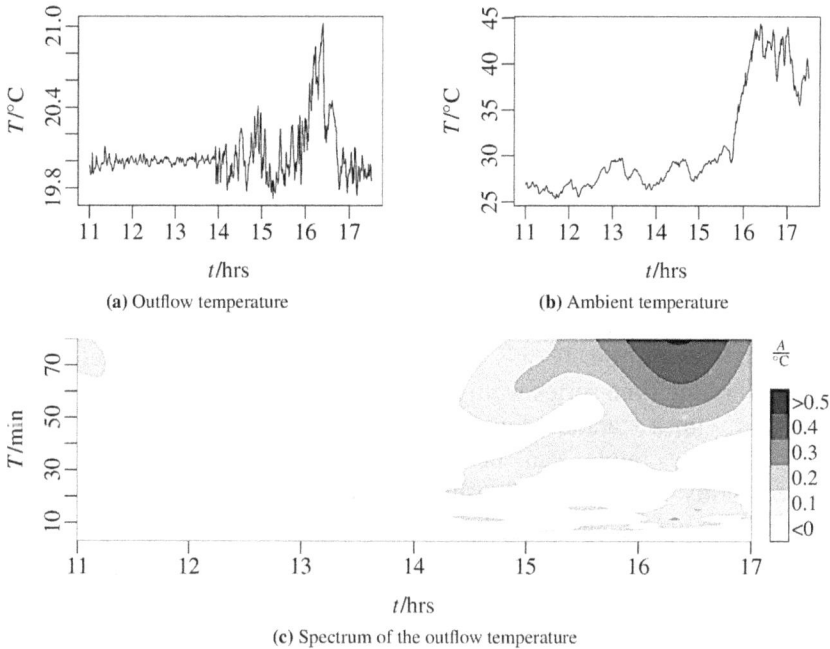

(a) Outflow temperature

(b) Ambient temperature

(c) Spectrum of the outflow temperature

Fig. 2.19. Outflow temperature of a 300 kW cooling system. Panel (a) shows the performance of the outflow temperature. At about 4 p.m. the system reaches its control limits, because the ambient temperature (b) exceeds significantly the set-point (20 °C) of the outflow temperature. The high eigenfrequencies of the hydraulic system become visible from about 2:30 p.m. on.

Finally, the following code shows how to invert the frequencies to periods. Note that a re-sampling of the matrix must be performed, because the unequally spaced vector $1/f$ now maps to the corresponding spectrum. This task is done by the `plot.fft()` function, which also plots objects with the attribute `mode = "waterfall"`.

```
# calculating the waterfall diagram
wf <- waterfall(temp,time,nf = 3)
# change to periods = 1/f
wf$fx <- 1/wf$fx
# avoid "infinite" values
c <- !is.infinite(wf$fx)
wf$fx[1] <- 2 * max( wf$fx[c],na.rm = T )

# plot graph with hour-scale on the x-axis
wf$x <- wf$x/60

plot(wf, xlim = c(11,17), ylim = c(3,80), zlim =
c(0,0.5) )
```

2.4.8. Fragmented and Irregularly Sampled Data

Sometimes the nature of measurement prevents an equally spaced sampling. Observations of astrophysical bodies like the sun or the moon are examples of objects only visible at a certain daytime, therefore the corresponding time series would be fragmented, which makes a spectral analysis with a standard Fourier transform almost impossible. A nice application example can be found in "The Lick Planet Search: Detectability and Mass Thresholds" by A. Cumming [23]. Another demand would be the measurement with randomly sampled data, for instance the occurrence of an event like the appearance of a malfunctioning product in a production lane, or even measurements with high jitter.

In a general point of view, when irregular sampling takes place, the time instances t_n depend on the sample number n, thereby becoming unique for each individual sample taken. In particular the time interval

$$\Delta t = t_{n-1} - t_n \neq \text{const}$$

is not constant anymore, so the sampling time becomes an additional data vector. This enables one advantage in case of randomly sampled data. Given a stationary process, it now becomes possible to estimate signal amplitudes for frequencies which are in the order of $f_{max} \approx \mathcal{O}\left(\min(\Delta t)^{-1}\right)$, with the maximum frequency corresponding to the minimal distance between two points. This is a bit in contrast to the average Nyquist frequency $\overline{f}_c$, where

$$f_{max} > \frac{1}{2} \frac{N}{\max(t) - \min(t)} = \overline{f}_c, \qquad (2.113)$$

which can be defined by the time range of the time series and the total number of samples N.

Let us introduce the Lomb-Scargle periodogram as a consistent estimator for spectral properties of discrete data. The estimation of the amplitude spectra in that sense is not a mathematical transform like the Fourier transform, but rather a statistical measure about the input data. A main difference compared to the DFT are the frequencies of interest, which can be selected freely to analyze a data set for amplitudes and phases regarding multiple frequencies. In addition to that a "False-Alarm-Probability" (FAP) describes the probability that a detected peak might be the result of noise. The associated mathematical details of this statistical approach can be found in the papers by A. Mathias [24], K. Hocke [25], J. Scargle [8] and N. Lomb [26], who originally developed the method.

In principle, the concept of the procedure is about a least squares optimization of the parameters A_i and B_i for the model function

$$y(t) = \sum_i A_i \cdot \cos\left(\omega_i \cdot (t - \tau_i)\right) + B_i \cdot \sin\left(\omega_i \cdot (t - \tau_i)\right), \quad (2.114)$$

which is fitted to the data. The signal $y(t)$ is represented as a sum of cos- and sin-functions with individual circular frequencies ω_i and an additional phase shift τ_i. Remember, the introductory Section 2.2.2 shows a slightly different approach, while utilizing the already introduced trigonometric identities (2.4) and (2.5). However, the traditional calculation of a Newton algorithm for a non-linear model would take several iteration steps until the result converges. Alternatively, the optimal parameters can be estimated by one single matrix inversion, see [24], which is enabled by properly considering the orthogonality property of the trigonometric functions. The mathematical proof can be found in "A generalized inverse for matrices" by R. Penrose [27]. As a necessary condition the relation

$$\sum_{n=1}^{N} \cos\left(\omega_i \cdot (t_n - \tau_i)\right) \cdot \sin\left(\omega_i \cdot (t_n - \tau_i)\right) = 0 \qquad (2.115)$$

must be fulfilled to obtain an optimal result for A_i and B_i. Expression

$$\tan\left(2\omega_i \cdot \tau_i\right) = \frac{\sum_n \sin(2\omega_i \cdot t_n)}{\sum_n \cos(2\omega_i \cdot t_n)} \tag{2.116}$$

can be deduced directly from (2.115) to determine the optimal shift τ_i. However, the procedure is very similar to Eq. (2.8) from Section 2.2.2, but now a model function with two individual trigonometric terms is investigated. Hereby, the free parameter τ_i effect two things: First, it minimizes the truncation error ε (see equation (2.17)), which originates from the misfit of integer period length to time intervals of arbitrary lengths. Thus, the parameter τ_i shifts the kernel functions, so that either $\int \cos\left(\omega_i\left(t - \tau_i\right)\right) \, dt = 0$ or the corresponding sin-integral become zero. Second, it enables condition (2.115), which will select either the cos-term or the sin-term of the model function for calculation. This can be seen when multiplying

$$\sum_n y(t_n) \cdot \cos\left(\omega_i \cdot \left(t_n - \tau_i\right)\right) = A_i \cdot \sum_n \cos^2\left(\omega_i \cdot \left(t_n - \tau_i\right)\right) \tag{2.117}$$

so that one of the two model terms is neglected. From (2.117) follows

$$A_i = \frac{\sum_n y(t_n) \cdot \cos\left(\omega_i \cdot \left(t_n - \tau_i\right)\right)}{\sum_n \cos^2\left(\omega_i \cdot \left(t_n - \tau_i\right)\right)} \quad \text{and} \quad B_i = \frac{\sum_n y(t_n) \cdot \sin\left(\omega_i \cdot \left(t_n - \tau_i\right)\right)}{\sum_n \sin^2\left(\omega_i \cdot \left(t_n - \tau_i\right)\right)}$$

so that the sinusoidal amplitude can be expressed by $\sqrt{A_i^2 + B_i^2}$.

Next to that, the parameters R, I, C and S will be defined as follows:

$$R\left(\omega_i\right) \equiv \sum_n y_n \cos\left(\omega_i \cdot \left(t_n - \tau_i\right)\right) \tag{2.118}$$

$$I\left(\omega_i\right) \equiv \sum_n y_n \sin\left(\omega_i \cdot \left(t_n - \tau_i\right)\right) \tag{2.119}$$

$$C\left(\omega_i\right) \equiv \sum_n \cos^2\left(\omega_i \cdot \left(t_n - \tau_i\right)\right) \tag{2.120}$$

$$S(\omega_i) \equiv \sum_n \sin^2\big(\omega_i \cdot (t_n - \tau_i)\big) \tag{2.121}$$

Finally, the normalized power spectral density P, the absolute amplitude A and the phase φ

$$P(\omega_i) = \frac{1}{2\sigma^2}\left(\frac{R(\omega_i)^2}{C(\omega_i)} + \frac{I(\omega_i)^2}{S(\omega_i)}\right) \tag{2.122}$$

$$A(\omega_i) = \sqrt{\underbrace{\left(\frac{R(\omega_i)}{C(\omega_i)}\right)^2}_{A_i} + \underbrace{\left(\frac{I(\omega_i)}{S(\omega_i)}\right)^2}_{B_i}} \tag{2.123}$$

$$\varphi(\omega_i) = -\left(\arctan\left(\frac{I}{R}\right) + \omega_i\tau_i\right) \tag{2.124}$$

can be calculated from these coefficients. Here σ describes the standard deviation of the discrete data vector y_n. In comparison to Eq. (2.17) from Section 2.2.2, the equations above work quite similarly. But now the modified function argument – extended by τ_i – and the correction terms $C(\omega_i)$ and $S(\omega_i)$ lead to the optimal least square fit solution.

In the limit of an infinite number of equally spaced samples, the Lomb-Scargle estimator converges to the Fourier transform, so it becomes a consistent estimator [24]. The last statement is very important, because with an increasing number of samples the error reduces until the result converges to the true amplitude and phase.

To value the significance of the estimated amplitude the "false alarm probability" (FAP)

$$p(P < P_0) = 1 - \left(1 - e^{-P}\right)^M \tag{2.125}$$

is defined. Here the probability p, that there is no other larger amplitude P than P_0, is expressed in terms of the exponential function above. For small values of p the approximation

$$p(P < P_0) \approx M \cdot e^{-P}\Big|_{p \ll 1} \tag{2.126}$$

can be used. The free parameter M counts the *independent* frequencies in the data. These are difficult to measure *a priori*, but it turns out that $M = N/2$ achieves sufficient results. A brief discussion on this issue can be read in the work of R. Townsend [28, Chapter 6.2], M. Zechmeister [29] and A. Cumming [23].

Implementation. The calculation of the phase is tricky. Invoking the arctan2() function is mandatory. Nevertheless, to prevent errors the phase φ is calculated by

$$\varphi = -\left(\arctan2\left(\frac{I}{N}, \frac{R}{N} \right) + \omega_i \tau_i \right), \qquad (2.127)$$

which takes the normalized components I and R. This is necessary for the arctan2-implementation in R.

Next to that, if the suggestions made by R. Townsend [28] are taken into account then the above algorithm can be shortened. The problem is that the straightforward implementation runs *twice* over the whole data set, while calculating τ prior to the rest of the parameters. A keen refactoring of the equations above will minimize the computational cost. For a certain frequency ω_i the amplitudes

$$A(\omega_i) = \sqrt{\left(\frac{c_\tau \cdot XC + s_\tau \cdot XS}{c_\tau^2 \cdot CC + 2c_\tau \cdot s_\tau \cdot CS + s_\tau^2 \cdot SS} \right)^2 + \left(\frac{c_\tau \cdot XS - s_\tau \cdot XC}{c_\tau^2 \cdot SS + 2c_\tau \cdot s_\tau \cdot CS + s_\tau^2 \cdot CC} \right)^2}$$

$$(2.128)$$

can be calculated out of the parameters

$$XC = \sum_j y_j \cdot \cos\left(\omega_i \cdot t_j \right) \qquad (2.129)$$

$$XS = \sum_j y_j \cdot \sin\left(\omega_i \cdot t_j \right) \qquad (2.130)$$

$$CC = \sum_j \cos^2\left(\omega_i \cdot t_j \right) \qquad (2.131)$$

$$SS = \sum_j \sin^2\left(\omega_i \cdot t_j \right) \qquad (2.132)$$

$$CS = \sum_{j} \cos\left(\omega_i \cdot t_j\right) \cdot \sin\left(\omega_i \cdot t_j\right) \qquad (2.133)$$

$$\tau_{\mathrm{L}} = \arctan\left(\frac{2 \cdot CS}{2 \cdot CC - 1}\right) \qquad (2.134)$$

$$c_\tau = \cos\left(\omega \cdot \tau_{\mathrm{L}}\right) (2.135)$$

$$s_\tau = \sin\left(\omega_i \cdot \tau_{\mathrm{L}}\right) (2.136)$$

in one single loop. A vectorized *R*-code example is given below. Here the data and the corresponding frequencies form the matrix omega.x, which is processed successively.

```
# put everything in a matrix.
# One frequency per column.
# x correspondes to the time.
# y_ corresponds to the mean-free
# data vector.
omega.x <- x %*% t(omega)
co <- cos(omega.x); si <- sin(omega.x)
# use trigonometric identities
co2 <- co^2; si2 <- 1 - co2
si <- sqrt(1-co2)

CC <- colSums(co2)
SS <- colSums(si2)
YCS <- colSums(y_ * co)
YSS <- colSums(y_ * si)
CS <- colSums(si*co)
tauL <- atan(2 * CS / (2*CC - 1) / (2*omega) )
ct <- cos(omega * tauL)
st <- sin(omega * tauL)
ct2 <- ct^2
st2 <- 1-ct2

ctstCS <- 2 * ct * st * CS
R <- (ct*YCS + st*YSS)
I <- (ct*YSS - st*YCS)
C <- (ct2 * s[[3]] + ctstCS + st2 * s[[4]])
S <- (ct2 * s[[4]] - ctstCS + st2 * s[[3]])

# a trick to reduce numeric error
# in atan2()
l <- sqrt(R^2+I^2)
A <- sqrt( (R/C)^2 + (I/S)^2 )
phi <- - omega*tauL- atan2((I/l),(R/l))
```

The *R*-code illustrates how to use the specialties of the language. Instead of programming a for statement the faster vector and matrix operations of *R* can be invoked.

Application. To show the power of the method let us first assume a simple example. The function

$$y(x) = \sin(2\pi \cdot 7 \cdot x) \tag{2.137}$$

is going to be sampled with $N = 101$ equally spaced samples. Remember the last point of the data vector equals the first point of the period of the signal. As discussed in the application Section 2.4.1, the violation of the periodicity would lead to an error in the result of the Fourier transform. In addition to that, approximately 30 % of the data were deleted.

Subsequently, Fig. 2.20 shows the fragmented signal and the corresponding periodogram, which is calculated with the `spec.lomb(x, y, f)` function.

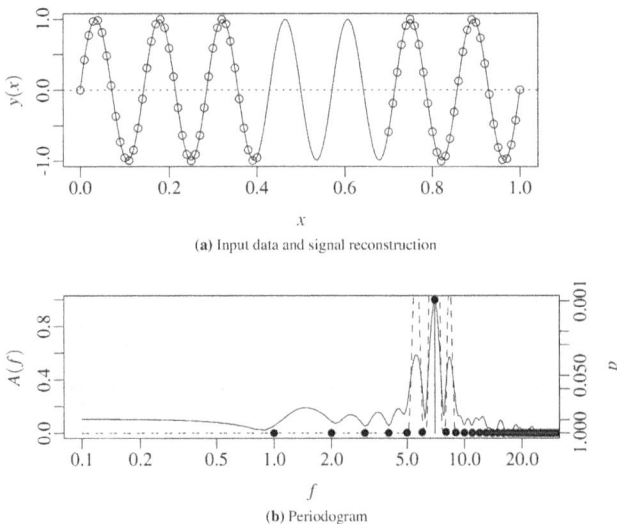

(a) Input data and signal reconstruction

(b) Periodogram

Fig. 2.20. Simple Lomb-Scargle periodogram. The open symbols in the top panel (a) display the input signal. The solid line shows the reconstruction out of the spectrum. Panel (b) focuses on the periodogram. Here, the solid line corresponds to the spectral amplitude and the black points indicate the results of a DFT, as if there were no gap in the data. The dashed line represents the false alarm probability *p*. Note, the value of 1 is located at the bottom, whereas smaller values are displayed above.

```
x <- seq(0, 1, by = 0.01)
x.new <- seq(0, 1, by = 1e-3)
yorg <- function(x)
return( sin( 2*pi * 7*x ) )
# create the gap
cond <- !(x > 0.4 & x < 0.7)
x <- x[cond]; y <- yorg(x)

l <- spec.lomb(x = x, y = y, f = seq(0, 25, by = 0.1) )
lf <- filter.lomb(l, newx = x.new
                  ,phase = "lin"
                  ,threshold = 3
                  )
```

The code above shows how Fig. 2.20 can be created. Note that the frequency vector f contains 250 different frequencies to analyze the data. Compared to the single sided spectrum of the *unfragmented* data's analytic signal representation (black dots in Fig. 2.20b) the Lomb-Scargle periodogram produces large side band amplitudes to the left and right of the main amplitude. This is quite typical if the data is non-uniformly sampled or even patchy. The dashed line represents the false alarm probability, which tends to zero if the corresponding amplitude is significant.

The *spectral* package also provides a filter.lomb() function, with which the most significant amplitudes can be extracted for reconstruction. Provided the continuous sampling vector x.new, the result is a new data set in which the remaining gaps are filled.

A more complex example is given in Fig. 2.21.

Here the function

$$y(x) = \sin(2\pi \cdot (x + \mathsf{N}\ (0, \Delta x / 4))) +$$

$$+ \sin(2\pi \cdot 20 \cdot (x + \mathsf{N}\ (0, \Delta x / 4))) + \mathsf{N}\ (0,1) \qquad (2.138)$$

consists of two sin-terms, the location of sampling jitters as well as the amplitude itself. In the simple test case here, normal distributed noise is assumed and the resulting data vectors x_n and y_n are $N = 101$ elements in length. However, for reconstruction the most significant amplitude values are selected again, so that the dashed line in panel (a) will fill the gaps correspondingly.

For further reading about non-uniformly sampled data the adaptive approach introduced by P. Stoika [30] is recommended. This method has a better signal to noise ratio, but will have much more computational cost.

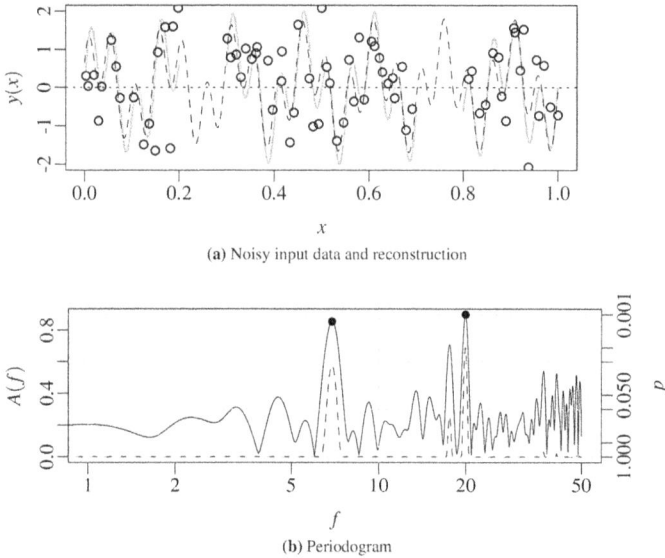

(a) Noisy input data and reconstruction

(b) Periodogram

Fig. 2.21. Randomly sampled data with noise and gaps. The bold gray line in (a) is the discontinuous signal. The circles represent the sampling points, which are overlaid with some strong noise and jitter. The dashed line represents the reconstruction. In (b) the corresponding Lomb-Scargle periodogram is given. The black line is the amplitude spectrum and the dashed line indicates p. The black points show the amplitude maxima, which the `filter.lomb()` function chose for reconstruction.

2.5. Conclusion

Today, spectral analysis is one of the key methods in data processing. It helps to filter signals, to accelerate calculations or to identify processes. While the complex mathematics is already documented in the given references and the common text books, the present work focuses on the *application* of the basic spectral methods in one dimension. Here working principles of the introduced methods and their possible misunderstandings are of interest. Small examples support the explanations about the work of operation and show the "dos and don'ts".

The idea of the present work is to show how the complex underlying theory of spectral methods can be applied to solve real world tasks. A dedicated tool supporting this intent is the *spectral* package in the statistic environment *R*, which was developed contemporaneously to this chapter.

References

[1]. R Core Team, The R Project for Statistical Computing, Vienna, Austria, 2016, https://www.r-project.org/

[2]. M. Seilmayer, spectral: Common Methods of Spectral Data Analysis, 2016, R package version 1.0, http://CRAN.R-project.org/package=spectral

[3]. M. Seilmayer, M. Ratajczak, A guide on spectral methods applied to discrete data in one dimension, *Journal of Applied Mathematics*, Vol. 2017, 2017, pp. 1-27.

[4]. S. W. Smith, The Scientist & Engineer's Guide to Digital Signal Processing, 1st Ed., *California Technical Pub*, San Diego, 1997.

[5]. L. Cohen, Time-Frequency Analysis, *Prentice Hall*, Englewood Cliffs, N. J, 1995.

[6]. I. N. Bronštejn, K. A. Semendjaev, Taschenbuch der Mathematik, 5th Ed., *Harri Deutsch*, Thun, 2001.

[7]. A. J. Jerri, The Shannon sampling theorem – Its various extensions and applications: A tutorial review, *Proceedings of the IEEE*, Vol. 65, Issue 11, 1977, pp. 1565-1596.

[8]. J. D. Scargle, Studies in astronomical time series analysis. II – Statistical aspects of spectral analysis of unevenly spaced data, *The Astrophysical Journal*, Vol. 263, December 1982, pp. 835-853.

[9]. N. E. Huang, Z. Shen, S. R. Long, M. C. Wu, H. H. Shih, Q. Zheng, N.-C. Yen, C. C. Tung, H. H. Liu, The empirical mode decomposition and the Hilbert spectrum for nonlinear and non-stationary time series analysis, *Proceedings of the Mathematical, Physical and Engineering Sciences*, Vol. 454, Issue 1971, March 1998, pp. 903-995.

[10]. D. E. Vakman, L. A. Vainshtein, Amplitude, phase, frequency – Fundamental concepts of oscillation theory, *Soviet Physics Uspekhi*, Vol. 20, Issue 12, December 1977, pp. 1002-1016.

[11]. J. Duoandikoetxea Zuazo, Fourier Analysis, Graduate Studies in Mathematics, Vol. 29, *American Mathematical Society*, Providence, R. I., 2001.

[12]. R. Hoffmann, Signalanalyse und -erkennung: Eine Einführung für Informationstechniker, *Springer*, Berlin, Heidelberg, 1998.

[13]. C. E. Shannon, Communication in the presence of noise, *Proceedings of the IRE*, Vol. 37, Issue 1, 1949, pp. 10-21.

[14]. J. M. Whittaker, Interpolatory Function Theory, *University Press*, Cambridge, Eng., 1935.

[15]. E. M. Stein, G. L. Weiss, Introduction to Fourier Analysis on Euclidean Spaces, Princeton Mathematical Series, Vol. 32, *Princeton University Press*, Princeton, N. J., 1975.

[16]. R. C. Gonzalez, R. E. Woods, Digital Image Processing, 2nd Ed., *Prentice Hall*, Upper Saddle River, N. J., 2002.

[17]. T. H. Cormen, C. Stein, C. E. Leiserson, R. L. Rivest., 2nd Ed., Introduction to Algorithms, *The MIT Press*, Cambridge, Mass., 2001.

[18]. M. Frigo, S. G. Johnson, The design and implementation of FFTW3, *Proceedings of the IEEE*, Vol. 93, Issue 2, 2005, pp. 216-231.

[19]. L. Debnath, F. A. Shah, Wavelet Transforms and Their Applications, *Birkhäuser Boston*, M. A., 2015.

[20]. R. J. Marks, Handbook of Fourier Analysis & Its Applications, Vol. 800, *University Press Oxford*, C. T., 2009.

[21]. N. Wiener, Extrapolation, Interpolation, and Smoothing of Stationary Time Series, Vol. 2, *MIT Press*, Cambridge, M. A., 1949.

[22]. M. I. Skolnik, Introduction to Radar Systems, 2nd Ed., *McGraw-Hill*, New York, 1980.

[23]. A. Cumming, G. W. Marcy, R. P. Butler, The lick planet search: detectability and mass thresholds, *The Astrophysical Journal*, Vol. 526, Issue 2, 1999, 890.

[24]. A. Mathias, F. Grond, R. Guardans, D. Seese, M. Canela, H. H. Diebner, G. Baiocchi, Algorithms for spectral analysis of irregularly sampled time series, *Journal of Statistical Software*, Vol. 11, Issue 2, 2004, pp. 1-30.

[25]. K. Hocke, N. Kämpfer, Gap filling and noise reduction of unevenly sampled data by means of the Lomb-Scargle periodogram, *Atmospheric Chemistry and Physics*, Vol. 9, Issue 12, 2009, pp. 4197-4206.

[26]. N. R. Lomb, Least-squares frequency analysis of unequally spaced data, *Astrophysics and Space Science*, Vol. 39, February 1976, pp. 447-462.

[27]. R. Penrose, J. A. Todd, A generalized inverse for matrices, *Mathematical Proceedings of the Cambridge Philosophical Society*, Vol. 51, Issue 3, July 1955, pp. 406-413.

[28]. R. H. D. Townsend, Fast calculation of the lomb-scargle periodogram using graphics processing units, *The Astrophysical Journal Supplement Series*, Vol. 191, Issue 2, December 2010, pp. 247-253.

[29]. M. Zechmeister, M. Kürster, The generalised Lomb-Scargle periodogram. A new formalism for the floating-mean and Keplerian periodograms, *Astronomy & Astrophysics*, Vol. 496, Issue 2, March 2009, pp. 577-584.

[30]. P. Stoica, J. Li, H. He, Spectral analysis of nonuniformly sampled data: A new approach versus the periodogram, *IEEE Transactions on Signal Processing*, Vol. 57, Issue 3, March 2009, pp. 843-858.

Chapter 3

Mathematical Tools for Measurements. Application for Quality Control Based Material Testing and Characterization

Salah Bouhouche

3.1. Quality Assurance and Conformity Declaration

From raw material until the final product, quality control and defect detection need an online monitoring and quality evaluation system. Sometimes this is a difficult task in practice because continuous measurements are affected by several factors such as physical constraints, maintenance cost, or non-existence of reliable technology. The quality control services assume a hard responsibility to guarantee a product or service free or with acceptable defects. The acceptability of defect is defined by a target range of the quality index variability.

As defined in different standards of quality management system and also in the good practice guidelines, the quality control services must be independent of the production: The Independence and the impartiality conditions must be respected; therefore the organization and the management of different interactions between different sub systems must be achieved. This aspect of organization is important to be achieved before to implement the quality control methods.

Quality control is generally based on measurement of characteristics of the controlled product, the measurement can be achieved in online by required sensors and equipment, and in this case a huge data base is produced during the production process. A computerized quality control system is then necessary to implement, it processes the data and a quality

Salah Bouhouche
Research Center in Industrial Technologies CRTI, Algiers, Algeria

assessment is given for decision making. There are many successful use of methods based on modeling and simulation [1-4], however, they need improvement particularly in the case of excessive measurement uncertainty.

In the case of the semi continuous or discontinuous production processes which are sometimes called serial production, the measurement is not made continuously, and a sampling plan is then applied to achieve an acceptable quality control task. A random sampling technique is applied, a sampling standard is generally recommended according to the type of products and processes. By the same manner of the continuous measurement process, the sampled data are processed using methods and models.

Fig. 3.1 shows the principle of quality control chart in the case of continuous measurement and sampling.

Procedural approach is generally applied in the case when a simple control without a complex computing techniques are not needed, this is applicable for a typical applications. However, the quality control of complex system needs an implementation of complex procedures based models.

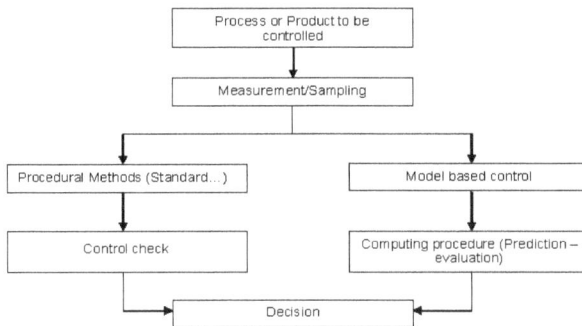

Fig. 3.1. Principal of quality control methods.

There are many works and publications in the field of quality control, most of them are shown its limits particularly in the case of noisy data, for that a particular importance should be given to the conformity assessment with uncertainty of measurement, this last is generated by different factors. As shown in the following Fig. 3.2, the product characterized by the measurement points 1, 2 and 3 are considered as conform because the measurement values associated with the uncertainty remain under the control limit. The measurement point 4 is in the limit of acceptability, by considering the uncertainty of measurement, this product becomes not acceptable. The measurement point 5 present a case of the inacceptable product because the measurement point without uncertainty of measurement is out of the control limit. According to the importance of uncertainty, a product conform can be considered as non-conform, this is clearly shown in the measurement point 6. The uncertainty of measurement should be accurately computed for a good conformity declaration.

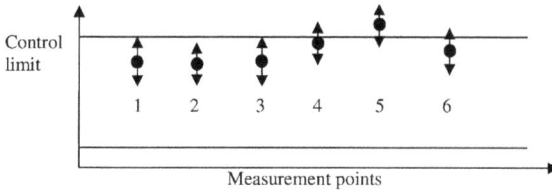

Fig. 3.2. Principle of quality control and conformity assessment chart.

The principle of quality control given in Fig. 3.2 is simple and do not consider the prediction based modeling, such approach remains limited particularly in the case of unstable state of noise. This aspect is generally based on the procedural approach with a limited computing procedure. For a more accurate quality control system, the uncertainty value should be correctly estimated using the required models and methods. This approach is based on the advanced modeling tools and its allied techniques such as intelligent, data mining and complex models, also correlation analysis should be studied to assess the quality influencing factors.

In material testing and characterization many analysis and tests are needed, the measured parameters of samples are given by X_i and the global characterization is formulated by a global formalism as given by Fig. 3.3 below. The model output of the global model is given by y_i.

Let the complex relationship between analysis given by the following equation:

$$y_i = F_i(X_1, X_2, ..., X_n) \tag{3.1}$$

The complex relationship F_i is obtained using model identification method based on Least Square Algorithm and its allied forms [5-7].

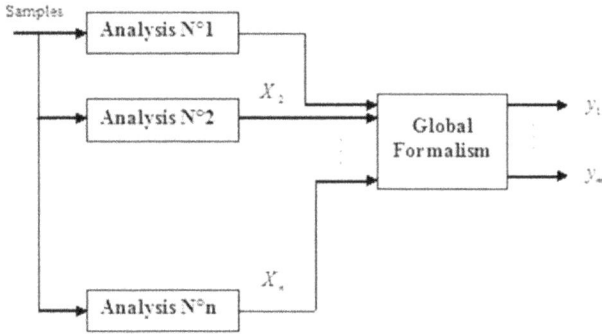

Fig. 3.3. Model structure of the global characterization.

Each measured parameter X_i is perturbed by a measurement error ΔX_i, then each input (i) becomes $(X_i + \Delta X_i)$, the measurement perturbation ΔX_i is characterized by its statistical properties such the distribution function and its associated parameters such as mean, standard deviation and so. Those perturbations are propagated along the global characterization model and give a global distribution model shown in Fig. 3.4.

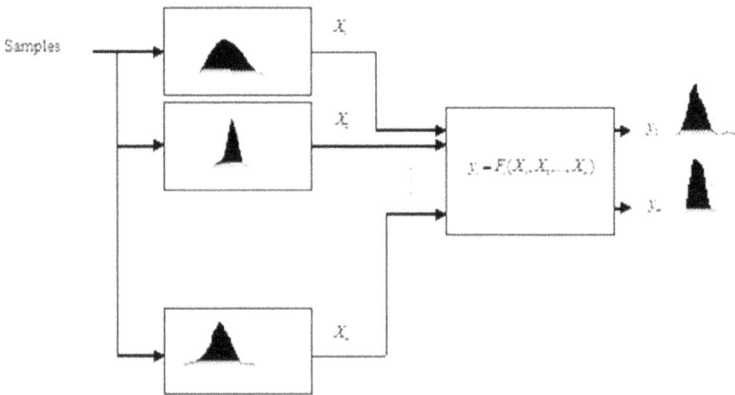

Fig. 3.4. Model of the propagation of distributions.

3.2. Overview of Mathematical Tools in Measurement

Measurement results are widely used along our daily activities, it gives results on which very important decisions can be made for conformity assessment for example, and therefore, such measurement must be more accurate. Because a measurement results cannot be a full deterministic phenomenon, it is very important to estimate the interval of the random part which is determinant for decision-making. Repetitive activities generate results not fully equal measurement, trends analysis using mathematical tools is necessary, the complexity of such tools depend on the static and dynamic changes of observed measurements. Four (04) sub domains can be considered as follows:

Descriptive statistics:

Descriptive statistics is generally used to describe the evolution of measurement using some descriptive indication such as mean, standard deviation, distribution function etc.

Simple correlation:

Simple correlation is used to define a relation between two variables generally characterizing the measured as a function of an independent variable, this is used in indirect measurement for example the measured temperature is computed by the variation of the electrical resistance (in the case of PT100). Fig. 3.5 gives the principle of measurement function.

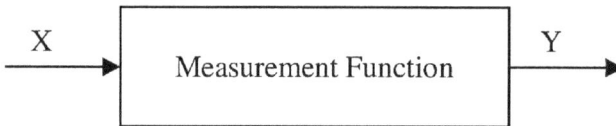

Fig. 3.5. Typical model of measurement.

$$y = f(x),\qquad(3.2)$$

$$T(^{\circ}C) = f(R) = R_0 + a*R + b*R^2\qquad(3.3)$$

Multivariate models:

In the case of multivariate model, the model output i.e. the measured is expressed as a multivariate function of inputs (See Fig. 3.6):

$$y = f(x_1, x_2, ..., x_n) \qquad (3.4)$$

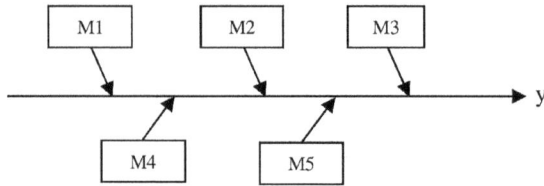

Fig. 3.6. Model of 5M influencing factors.

$$M_i = g_i(x_{Mi1}, x_{Mi2}, ..., x_{Min}) \qquad (3.5)$$

$$X_{in} = [x_{Mi1}, x_{Mi2}, ..., x_{Min}] \qquad (3.6)$$

Fig. 3.7 gives detaills of model interactions.

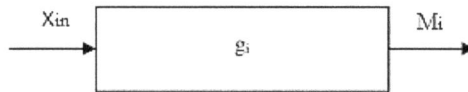

Fig. 3.7. Model details of influencing factors.

3.2.1. Structure and Model of Measurement Processes

The structure of measurement process is given in Fig. 3.8.

The measurement process generally gives repetitive results, the generated data must be analyzed using mathematical tools, and the descriptive data is analyzed using descriptive method. Table 3.1 gives the used typical measurement matrix.

The index "j" defines the measurement in repeatability; there is any change in the measurement conditions i.e. the measurement is made in the same conditions, the index "i" defines the measurement in the reproducibility conditions. There is at least one change in the conditions of measurement; for example we change from one test to another: The operator, the method, the medium, the material etc.

Using the measurement results given in the table, it is obtained a matrix of measurement having elementary elements x_{ij}.

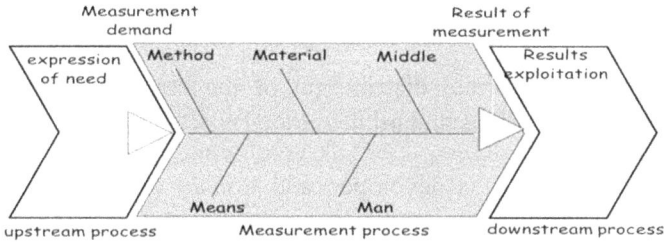

Fig. 3.8. Structure of measurement process.

Table 3.1. Typical measurement matrix.

Measurement in reproducibility conditions	Measurement in repeatability conditions			
	$j=1$	$j=2$	...	$j=n$
$i = 1$	x_{11}	x_{12}	...	x_{1n}
$i = 2$	x_{21}	x_{22}	...	x_{2n}
...	...	...	...	...
$i = m$	x_{m1}	x_{m2}	...	x_{mn}

Characterization of measurement precision is based on the analysis of the measurement matrix using conventional method such descriptive statistics; more details can be found particularly in different series of ISO 5725 standard.

3.2.2. General Formalism

An ideal process model can be defined by a complex relationship between its input (input space: χ) and output variables (output space: γ). This relationship is defined by the function Φ.

$$y = \Phi(\chi) \tag{3.7}$$

This latter, can be approximated by a model defining the dependency between an input data (X^t) and the output data (y^t).

$$y^t = f(X^t) + \varepsilon^t, \tag{3.8}$$

where $\varepsilon^t \in N(0, \sigma)$ and t is the measurement point, t-1, t-2 … are the past measurement points.

ε^t is a zero-mean normal distribution of the error term obtained in modeling step and σ its standard deviation.

Firstly, the model f is developed and trained using historical data. $D^{Hist} = (X^{Hist}, y^{Hist})$; where D^{Hist} is the historical domain containing the input/output data base. The estimated output at each measurement point or time (t) is given by:

$$\hat{y}_t = f(X^t), \tag{3.9}$$

where X^t is the input of the model depending on the model complexity. The input is composed by the input data $[x_t, ..., x_{t-Nx}]$, past values of the target data $[y_{t-1}, ..., y_{t-Ny-1}]$, and the prediction residual $[\varepsilon_{t-1}, ..., \varepsilon_{t-N\varepsilon-1}]$.

Using $[\varepsilon_{t-1}, ..., \varepsilon_{t-N\varepsilon-1}]$ as an input of the model can lead to a problem with the stability of the model because this term is used as a feedback. The model parameters are obtained by online identification methods using an adaptive procedure– based moving windows techniques. In some cases, characterized by a limited noise, the term $[\varepsilon_{t-1}, ..., \varepsilon_{t-N\varepsilon-1}]$ is cancelled.

For avoiding a possible loss of performance that may occur as a consequence of the model changes, an iterative form is necessary. In our case, iterative form based Levenberg-Marquardt Algorithm is used in the learning step of the neural network model.

The modeling objective is defined as:

$$\min_f \varepsilon_t^2 = \min_f (\hat{y}_t - y^t)^2 \tag{3.10}$$

The minimization of the criteria given by Eq.(3.6) is obtained using the least square methods existing in many references in this field, it exist simple LS that requires a matrix inversion, however the more popular is the RLS form which overcome matrix inversion. Because generally the measurement systems are multivariate, a multivariable version of RLS is developed to be applied to dynamic measurement system. Also, it is possible to consider simultaneous model parameters and state estimation based RLS [7-8].

Generally a multivariable measurement system is defined by a multivariable system with MISO:

$$y^t = f(x_1^t, x_2^t, ..., x_n^t) \tag{3.11}$$

For a linear system, Eq.(3.6) becomes:

$$y^t = b_1 x_1^t + b_2 x_2^t ... + b_n x_n^t \tag{3.12}$$

Define the parameter vector θ and the known information vector $\varphi(k)$ as:

$$\theta = [b_1, ..., b_n] \in R^n \tag{3.13}$$

$$\phi_t = [x_1^t, x_2^t, ..., x_n^t] \in R^n \tag{3.14}$$

From Eq. (3.13) and Eq. (3.14), we get the following identification model

$$y^t = \theta \phi_t^T + \varepsilon_t \tag{3.15}$$

$$\hat{y}^t = \theta \phi_t^T \tag{3.16}$$

Let $i = 1,2,...$ be an iterative step, L be the data length and θ_i an estimation of θ at the iteration i given by the minimization of the following criteria:

$$J(\theta) = \min(\sum_{i=1}^{L} [y^i - \theta_i \phi_i^T]^2) \Rightarrow \theta = \hat{\theta} \tag{3.17}$$

Minimizing the criterion function $J(\theta)$, the least square estimate of $\theta = \hat{\theta}$ for global time L_k and at every iteration i is obtained as follows:

$$\hat{\theta} = \left[\sum_{i=1}^{L} \phi_i \phi_i^T \right]^{-1} \left[\sum_{i=1}^{L} \phi_i y^i \right] \tag{3.18}$$

Eq. (3.18) can be written for every iteration k as:

$$\hat{\theta}_k = \left[\sum_{j=1}^{k} \phi_j \phi_j^T \right]^{-1} \left[\sum_{j=1}^{k-1} \phi_j y^j + \phi_k y^k \right] \tag{3.19}$$

Eq. (3.19) can be written for every time k as:

$$\hat{\theta}_k = \left[\sum_{j=1}^{k} \phi_j \phi_j^T \right]^{-1} \left(\left[\sum_{j=1}^{k-1} \phi_j \phi_j^T \right] \hat{\theta}_{k-1} + \phi_k y_k \right) \tag{3.20}$$

$$\hat{\theta}_k = \left[\sum_{j=1}^{k} \phi_j \phi_j^T \right]^{-1} \left(\left[\sum_{j=1}^{k} \phi_j y_j - \phi_k \phi_k^T \right] \hat{\theta}_{k-1} + \phi_k y_k \right) \tag{3.21}$$

$$\hat{\theta}_k = \hat{\theta}_{k-1} + \left[\sum_{j=1}^{k} \phi_j \phi_j^T \right]^{-1} \left(\phi_k y_k - \phi_k \phi_k^T \hat{\theta}_{k-1} \right) \tag{3.22}$$

Note that:

$$P_k = \left[\sum_{j=1}^{k} \phi_j \phi_j^T \right]^{-1} \tag{3.23}$$

Eq. (3.22) becomes:

$$\hat{\theta}_k = \hat{\theta}_{k-1} + K_k \left(y_k - \phi^T_k \hat{\theta}_{k-1} \right) \tag{3.24}$$

with

$$K_k = P_k \phi_k \tag{3.25}$$

Based on the matrix inversion Lemma:

$$(A + BCD)^{-1} = A^{-1} - A^{-1}B(DA^{-1}B + C^{-1})DA^{-1} \tag{3.26}$$

$$P_k^{-1} = P_{k-1}^{-1} + \phi_k \phi_k^T \tag{3.27}$$

We get the following recursive form of P_k:

$$P_k = P_{k-1} - \frac{P_{k-1}\phi_k\phi_k^T P_{k-1}}{1+\phi_k P_{k-1}\phi_k} \tag{3.28}$$

The developed RLS algorithm in its multivariable from is implemented using the following algorithm.

Algorithm N°1:

```
Initialization,

Define the model structure and set different variables:

N, Lk, IN is an Identity matrix, δ,

θ₁ = [0 0 ......0]N,

P = IN.10⁶,

For k = N:Lk,

        Acquire the inputs/output data,

        Compute the observation vector φk using Eq. (3.14),

        Compute the next value of Pk using Eq. (3.23),

        Compute the proportional gain Kk using Eq. (3.25),

        Use the recursive and compute the next estimation of
        θ̂k using Eq. (3.24),

        Compute the next value of Pk using Eq. (3.28),

End k,
```

If $\hat{\theta}_k - \hat{\theta}_{k-1} < \delta$ Stop.

3.2.3. Extension to Dynamic Measurement System

A dynamic system of measurement is generally given by a dynamic model; the finite difference equations as a numerical form of differential equations are used to model such measurement process.

In the dynamic way, note that the measurement model is given in the following form:

$$y(k) = f(y(k-1),..., y(k-N_y),u(k),u(k-1),..., u(k-N_u)) + \varepsilon(k) \tag{3.29}$$

This latter, can be approximated by a model defining the dependency between an input data $[y(k-1), ..., y(k-N_y), u(k), u(k-1), ..., u(k-N_u)]$ and an output data $(y(k))$, where $\varepsilon(k) \in N(0, \sigma)$ is a zero-mean normal distribution of the error term obtained in modeling step and σ its standard deviation.

The estimated output at each discrete time (k) is given by:

$$\hat{y}(k) = f([y(k-1), ..., y(k-N_y), u(k), u(k-1), ..., u(k-N_u)]) \quad (3.30)$$

Generally a multivariable measurement system is defined by a multivariable system with MISO:

$$y(k) = f(y(k-1), ..., y(k-N_y), u_1(k-1), u_1(k-2), ...,$$
$$u_1(k-N_{u1}), ..., u_p(k-1), ...u_p(k-N_{up}) + e(k) \quad (3.31)$$

For a linear system, Eq. (3.6) becomes:

$$y(k) = a_1 y(k-1) + ... + a_{Ny} y(k-N_y) + b_{11} u_1(k-1) + b_{12} u_1(k-2) + ...$$
$$+ b_{1Nu1} u_1(k-N_{u1}) + ... + b_{p1} u_p(k-1) + ... + b_{pNup} u_p(k-N_{up}) + e(k)$$
$$(3.32)$$

Define the parameter vector θ and the known information vector $\varphi(k)$ as:

$$\theta = [a_1, ..., a_{Ny}, b_{11}, ..., b_{1Nu1}, ..., b_{p1}, ..., b_{pNup}] \in R^{Ny+Nu1+...Nup} \quad (3.33)$$

$$\phi(k) = [y(k-1), ..., y(k-N_y), u_1(k-1), ..., u_1(k-N_{u1}), ...,$$
$$u_p(k-1), ..., u_p(k-N_{up})] \in R^{Ny+Nu1+...Nup} \quad (3.34)$$

From Eq. (3.33) and Eq. (3.34), we get the following identification model

$$y(k) = \theta\phi^T(k) + \varepsilon(k) \quad (3.35)$$

$$\hat{y}(k) = \theta\phi^T(k) \quad (3.35a)$$

$$J(\theta) = \min(\sum_{k=1}^{L_k}[y(k) - \theta\phi^T(k)]^2) \Rightarrow \theta = \hat{\theta}$$

3.2.4. Model Based Neural Network Model

Advanced process and quality control and monitoring require accurate process models. The development of analytical models from the relevant physical and chemical knowledge, especially for complex systems with phase changes, can be too costly or even technically impossible. For such models, based mainly on the data production, operational data should be capitalized. Many systems and processes are characterized by a non-linear dynamic behavior. Then, they need non-linear models. Indeed, Neural Networks have been shown to be able to approximate continuous non-linearities and have been applied in modeling of non-linear and complex processes of which the complexity is due to the large number of network weights. In practice, many non-linear processes are approximated by reduced order and possibly linear models and which are clearly related to the underlying process characteristics. The model identification principle using NN is given by Fig. 3.9. A model structure is chosen, the input and the output variables are defined, the modeling residual or error is computed and used as a tool to adapt the model parameters w_{ij}^t by the means of the computing procedure which generally includes a recursive form, more details about this method can bed founded in different documents [9-11].

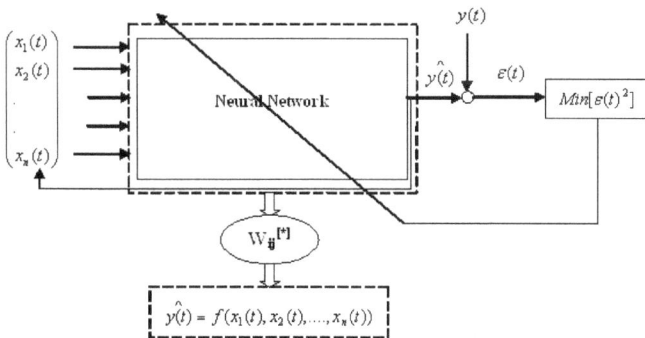

Fig. 3.9. Principle of breakout model identification using NN.

The estimated model output is defined by:

$$\hat{y}(k) = NN(\phi, w_{ij}^k), \tag{3.36}$$

where NN is a model structure, ϕ is the model input and w_{ij}^k are the NN weights. They are estimated by the corresponding algorithm that minimizes the modeling error as shown in Fig. 3.9. The recursive following form is taken from the Method of Levenberg-Marquardt which is a form of Newton family methods of the second order.

$$w_{ij}^k = w_{ij}^{k-1} - [H_{k-1} + \lambda_{k-1}I]^{-1} J_{k-1}, \tag{3.37}$$

where H_{k-1} and J_{k-1} are the Hessian and the Jacobean respectively λ_{k-1} is a constant defining the gradient and I is an identity matrix, the modelling error define as:

$$\varepsilon(k) = y(k) - \hat{y}(k) \tag{3.38}$$

Many published works have been developed in modeling and identification field using NN, the computing procedure is described by the following steps:

Step1: Initialize the network weights $w_{ij}^0 = [-0.5 \text{ to } +0.5]$,

Step2: Acquisition of inputs/outputs (x^t, y_t),

Step3: Compute the model output $\hat{y}_t$,

Step4: Compute the modeling error $\varepsilon(t) = y_t - \hat{y}_t$,

a) if $\varepsilon(t) \approx 0$, $w_{ij}^t = w_{ij}^{t-1} \to$ Stop: $w_{ij}^t = w_{ij}^{[*]}$,

b) Else, Adjust the NN Weights using the recursive Algorithm Eq. (3.37), Step5: Go to Step2.

3.2.5. Sensitivity Analysis

3.2.5.1. Sensitivity Based Derivative Form

The sensitivity analysis based derivative form is known as Taylor series, the model variation Δy is easily computed for simple function without high non linearity, but it remains limited for complex function, in this case numerical method such as Monte Carlo simulation is applied. For a multivariate function given by Eq. (3.11), the sensitivity is computed as follows [11-13]:

$$\Delta y = \sum_{j=1}^{m} \sum_{i=1}^{n} \frac{\partial^j f}{\partial x_i^j} \Delta x_i^j + \Delta x_i^j (0) \tag{3.39}$$

3.2.5.2. Sensitivity Based Modeland Monte Carlo Simulation [14-16]

The principle of the combined use of inferential model i.e. neural network model and Monte Carlo Simulation is given by the Fig. 3.10. The objective of the use of the inferential model is to take into account of different interactions between factors influencing the variability of measured parameters.

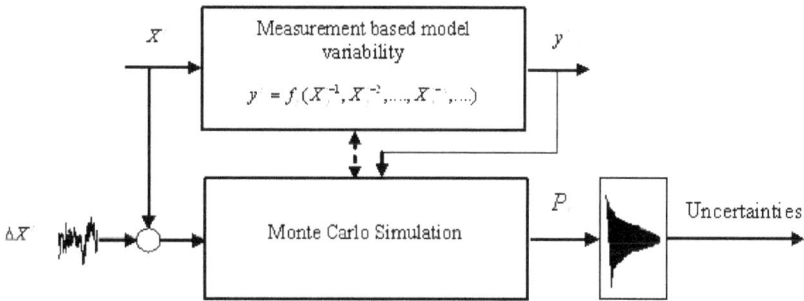

Fig. 3.10. Principle of the combined use of inferential model and Monte Carlo Simulation.

The evaluation method given by the above Fig. 3.10 is implemented as follows:

Step1: Acquisition of a new set of inputs/outputs (X^t, y_t) different of that used in the learning step (Section 3.2.2),

For $i = 1$ to N,

a): Perturbation of the input by adding a random value $[\Delta X^t]_i$ at each iteration (i), the input becomes $[X^t \pm \Delta X^t]_i$,

b): Compute the model output $\hat{y}_t$ using the neural network model obtained in the Section 3.2.2, c): Compute the modeling error $\varepsilon' = y_t - \hat{y}_t$,

d): Affect $\varepsilon' \to E_i'$ at each iteration (i),

end,

Step2: Statistical evaluation of the random matrix E_i': Analysis of the random distribution P_i,

Step3: Estimation of the global uncertainties from P_i i.e. use the a standard deviation,

Step4: End.

3.3. Application for Quality Control Based Material Testing and Characterization

3.3.1. Testing of Welding Quality

In this part, it is considered an application based measurement for testing of welding quality, about sixteen (3.16) points of measurement are chosen on area of thirty (3.30) welded sample as shown in Fig. 3.11. Hardness model and the homogeneity of chemical composition are considered.

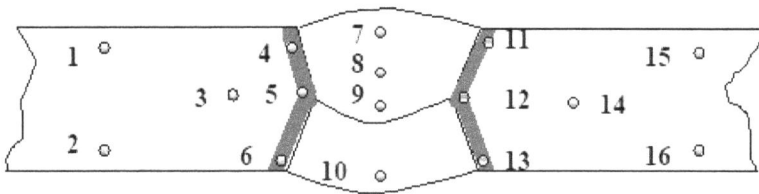

Fig. 3.11. Measurement points [17].

3.3.1.1. Hardness model

The followings measurements are obtained:

- Hardness (HV);

- Chemical composition such as (C, Mn, Si...): (NB+Ti+V) and C_{eq} is computed using the Carbone equivalent formula;

- Resilince (K_v);

- Traction (R_e).

Fig. 3.12 shows the measured data realized on thirty samples.

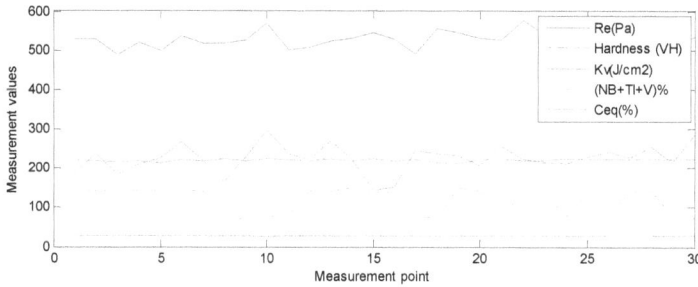

Fig. 3.12. Measurement data.

Firstly we are interesting to find correlation between all measured variables; the hardness is selected as an output of the model.

$$H = f(R_e, K_v, NB, C_{eq}) = \theta\phi^T, \qquad (3.40)$$

where $\theta = [a_1, a_2, a_3, a_4]$ and $\phi = [R_e, K_v, NB, C_{eq}]$.

By applying the identification algorithm given in the Section 3.2.2, the correlation function and errors are obtained and given in the followings figures (Figs. 3.13-3.15).

By the same manner the modeling procedure is applied to find relationship between hardness and chemical composition. Two models have been tested, the first is:

$$H = f(C, Mn, Si...) = \theta\phi^T, \qquad (3.41)$$

where $\theta = [a_1, a_2, a_3...]$ and $\phi = [C, Mn, Si...]$.

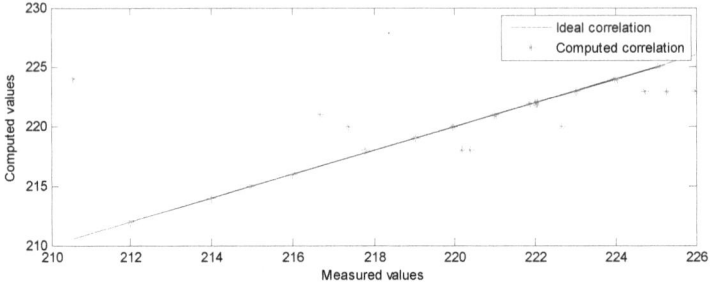

Fig. 3.13a. Correlation between measured and computed data of hardness.

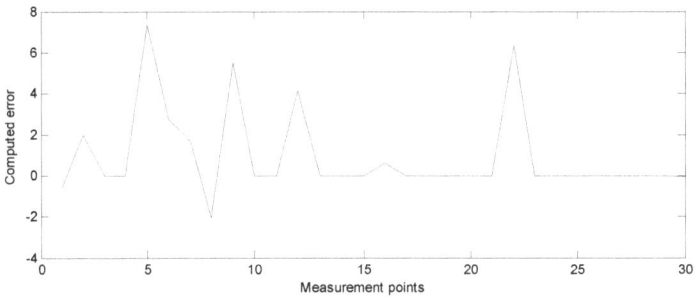

Fig. 3.13b. Computed error of hardness.

The correlation function between measured and computed data and the corresponding modeling error are given respectively by Figs. 3.14a and 3.14b.

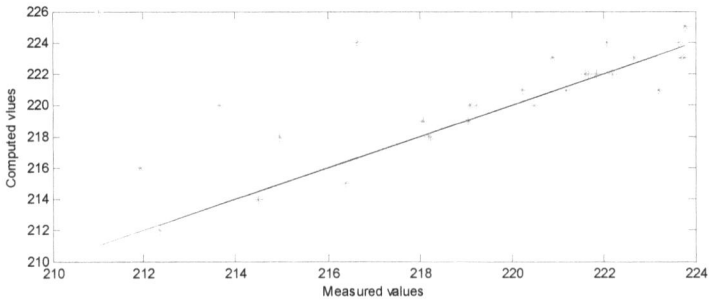

Fig. 3.14a. Correlation between measured and computed data of hardness –
All chemical analyzed elements.

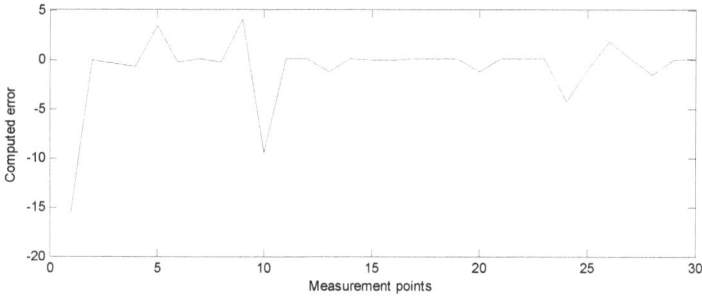

Fig. 3.14b. Computed error for Hardness – All chemical analyzed elements.

The followings figures showed the modeling results using a simple correlation variable between carbon content and Hardness.

$$H = f(C) \tag{3.42}$$

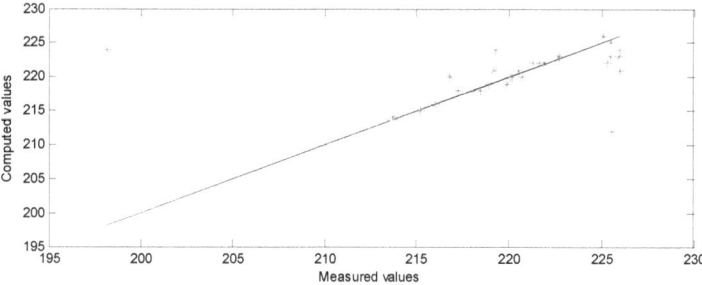

Fig. 3.15a. Correlation between measured and computed data of hardness – Carbone analyzed element.

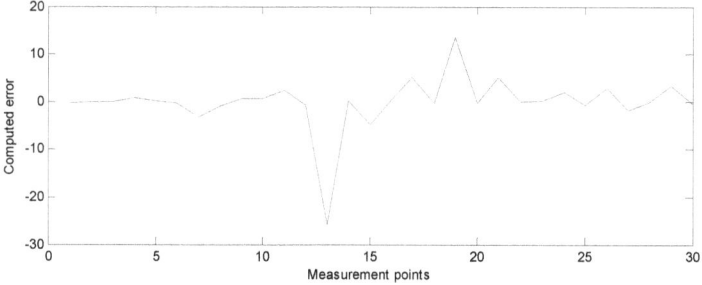

Fig. 3.15b. Computed error for Hardness – Carbone analyzed element.

139

3.3.1.2. Uncertainty Evaluation Based Sensitivity Analysis

In this part, the sensitivity analysis as a tool for uncertainty evaluation is obtained by the combined use of inferential model of hardness given by Eq. (3.40) and Monte Carlo simulation developed in the Section 3.2.5. The computing procedure of uncertainty based on Monte Carlo simulation is strongly recommended in this case because the derivative form cannot be easily obtained: The derivative form of neural network model is difficult to obtain, also the propagation of distributions cannot be easy considered in this case.

As shown in the followings figures, the propagation of uncertainty of inputs is computed using the above developed method; each input of the model is varied randomly using the corresponding distribution and its standard deviation. The influence on output i.e. the Hardness changes is obtained by an application of the Monte Carlo simulation loop.

Fig. 3.16 shows the variation of hardness according to the random change of input model in this case the Re (Fig. 3.16). The impact of different influencing factors taken in the model input on the hardness are given by Figs. 3.17-3.20 respectively for K_v model input, NB model input, C_{eq} model input and all model inputs together. More details and analysis will be developed in the next Section 3.3.2.

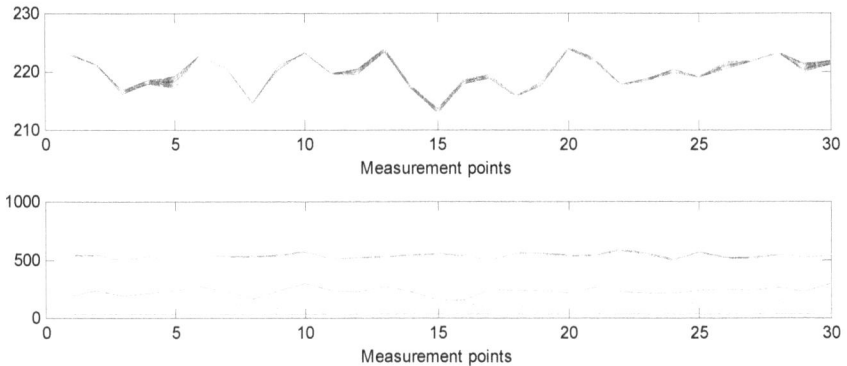

Fig. 3.16. Computed and measured data – R_e Input.

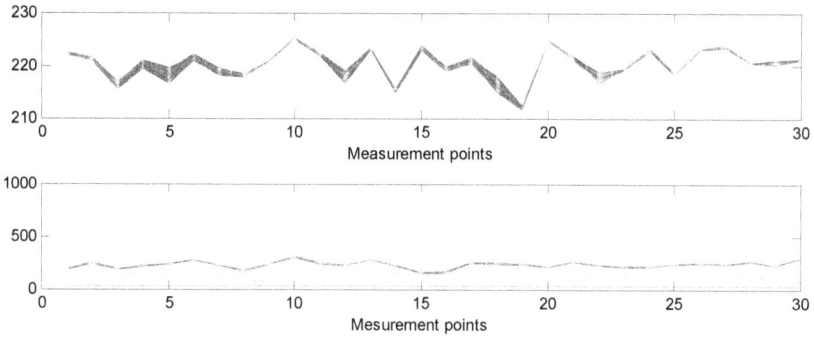

Fig. 3.17. Computed and measured data –K_v Input.

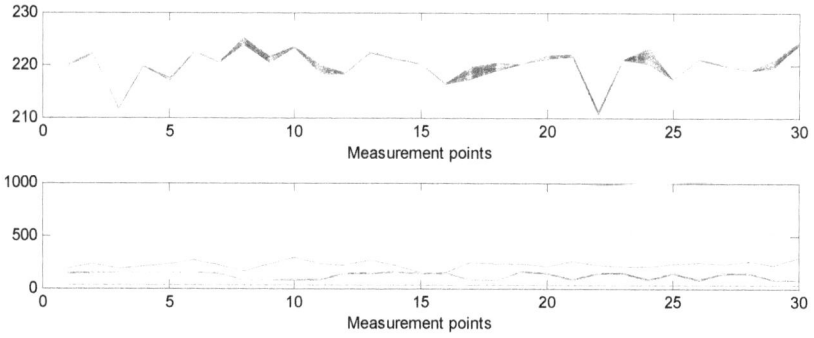

Fig. 3.18. Computed and measured data – NB Input.

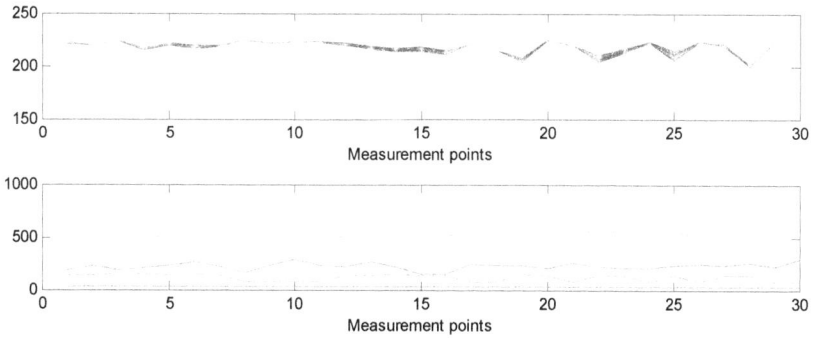

Fig. 3.19. Computed and measured data –C_{eq} Input.

141

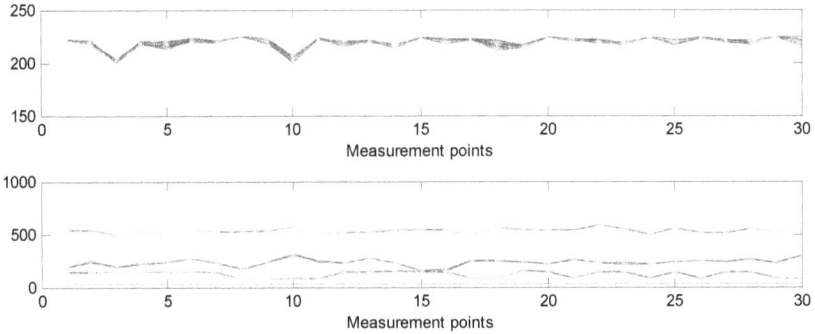

Fig. 3.20. Computed and measured data – All Model inputs.

3.3.2. Results

The results given in different curves are summarized in Tables 3.2 and 3.3 respectively for modeling precision and uncertainty evaluation. Table 3.2 shows the modeling precision for different models given by Eqs. (3.40)-(3.42). The Mean Squared Error (MSE) is used as index for evaluation, the most precise model is that given by Eq. (3.40) followed by models given by Eq. (3.41) and Eq. (3.42) respectively. The precision of the above cited model given by Eq. (3.41) is strongly correlated to the global composition of each measurement point, this is true because the hardness physics depend of chemical element than other parameters.

Table 3.2. Models precision – based MSE.

Model type	MSE
$H = f(R_e, K_v, NB, C_{eq})$	33.4341
$H = f(C, Mn, Si...)$	12.9870
$H = f(C)$	27.6786

Computed uncertainties are given in Table 3.3, Monte Carlo simulation coupled to the correlation model obtained using NN model are used for uncertainties evaluation, such approach is implemented numerically. The correlation model given by Eq. (3.40) has been validated using measurement data, all chemical and mechanical variables are taken into account, the key variables have been defined as follows:

- The chemical composition are given by (NB+V+Ti) and C_{eq},

- The mechanical variables are R_e and K_v.

Such model seem to be complete, but it remains imprecise comparatively to the model given by Eq. (3.41). The sensitivity of the model is computed for each input separately and for all variables together. The uncertainty value of each input is taken from the statistical variation; the standard deviation of each input distribution is computed and applied. The corresponding uncertainty of the output is computed by a hybrid procedure based on the NN model and Monte Carlo simulation. Random distribution of each input with the corresponding standard deviation is generated and its propagation via the NN model is obtained: The corresponding uncertainty is computed by the standard deviation of the obtained output distribution.

As shown in the Table 3.3, the uncertainty for all inputs together is less than the sum of the all uncertainty obtained using each input separately, this confirm the high non linearity of the Hardness model for which a NN model is applied.

Table 3.3. Estimated uncertainty.

Measurement model: $H = f(R_e, K_v, NB, C_{eq})$	
Uncertainty of model inputs[*]	**Uncertainty of model output**[*]
$u(R_e) = 2.63$, $u(K_v) = 0$, $u(NB) = 0$, $u(C_{eq}) = 0$	$u(H) = 0.59$
$u(K_v) = 4.49$, $u(R_e) = 0$, $u(NB) = 0$, $u(C_{eq}) = 0$	$u(H) = 0.948$
$u(NB) = 2.29$, $u(K_v) = 0$, $u(R_e) = 0$, $u(C_{eq}) = 0$	$u(H) = 0.8751$
$u(C_{eq}) = 0.083$, $u(NB) = 0$, $u(K_v) = 0$, $u(R_e) = 0$	$u(H) = 2.68$
$u(R_e) = 2.63$, $u(K_v) = 4.49$, $u(NB) = 2.29$, $u(C_{eq}) = 0.083$	$u(H) = 2.22$
[*] The uncertainty u() is expressed by the standard deviation values.	

3.4. Conclusion

A global formalism of advanced modeling and simulation has been developed, a computerized procedure for uncertainty evaluation based on a combined use of neural network model and Monte Carlo simulation has been considered. This formalism has been applied for quality control of welding, a set of data characterizing the measurement process on the welded sample has been made, different model structures have been tested and an optimal model has been retained, the MSE is used as an index for the model quality evaluation.

The retained model takes into account interactions between different inputs-output, the regression between chemical elements, mechanical data i.e. traction, resilience and hardness values in each point have been processed. Uncertainties have been evaluated using the above cited method.

References

[1]. M. D. Ma, J. W. Ko, S. J. Wang, M. F. Wu, S. S. Jang, S. S. Sheih, D. S. Wong, Development of adaptive soft sensor based on statistical identification of key variables, *Control Engineering Practice*, Vol. 17, 2009, pp. 1026-1034.

[2]. C. Zhao, T. Zhang, F. Wang, Nonlinear process monitoring based on kernel dissimilarity analysis, *Control Engineering Practice*, Vol. 17, 2009, pp. 221-230.

[3]. H. Wang, Improved extend Kalman particle filter based on Markov chain Monte Carlo for nonlinear state estimation, in *Proceedings of the International Conference on Image, Vision and Computing (ICIVC'12)*, Shanghai, China, 25-26 August 2012, pp. 1-7.

[3]. S. Bouhouche, M. Yahi, J. Bast, Combined use of principal component analysis and self organization map for condition monitoring in pickling process, *Applied Soft. Computing*, Vol. 11, 2011, pp. 3075-3082.

[4]. S. Bouhouche, Y. Laib, H. Sissaoui, J. Bast, Evaluation using online support vector machine and fuzzy reasoning. Application to condition monitoring of speeds rolling process, *Control Engineering Practice*, Vol. 18, Issue 9, 2010, pp. 1060-1068.

[5]. P. Kadlec, Review of adaptation mechanisms for data-driven soft sensors, *Computers and Chemical Engineering*, Vol. 35, 2011, pp. 1-24.

[6]. A. Zia, T. Kirubarajan, J. P. Reilly, D. Yee, An EM algorithm for nonlinear state estimation with model uncertainties, *IEEE Transactions on Signal Processing*, Vol. 56, Issue 3, 2008, pp. 921-936.

[7]. F. Ding, X. Liu, X. Ma, Kalman state filtering based least squares iterative parameter estimation for observer canonical state space systems using decomposition, *Journal of Computational and Applied Mathematics*, Vol. 301, 2016, pp. 135-143.

[8]. S. Pan, D. Xiao, S. Xing, S. S. Law, P. Du, Y. Li, A general extended Kalman filter for simultaneous estimation of system and unknown inputs, *Engineering Structures*, Vol. 109, 2016, pp. 85-98.

[9]. M. I. Ibrahimy, M. R. Ahsan, O. O. Khalifa, Design and optimization of Levenberg-Marquardt based neural network classifier for EMG signals to identify hand motions, *Measurement Science Review*, Vol. 13, Issue. 3, 2013, pp. 142-151.

[10]. Ö. Çelik, A. Teke, H. B. Yıldırım, The optimized artificial neural network model with Levenberg-Marquardt algorithm for global solar radiation estimation in Eastern Mediterranean Region of Turkey, *Journal of Cleaner Production*, Vol. 116, 2016, pp. 1-12.

[11]. U. Okkan, Application of Levenberg-Marquardt optimization algorithm based multilayer neural networks for hydrological time series modeling, *An International Journal of Optimization and Control: Theories & Applications*, Vol. 1, Issue 1, 2011, pp. 53-63.

[12]. Evaluation of Measurement Data – Guide to the Expression of Uncertainty in Measurement, JCGM 100:2008(F) GUM 1995, *JCGM*, 2008.

[13]. Evaluation of Measurement Data – Supplement 2 to the 'Guide to the Expression of Uncertainty in Measurement' – Extension to Any Number of Output Quantities, JCGM 102:2011, *JCGM*, 2011.

[14]. Evaluation of Measurement Data– Supplement 1 to the 'Guide to the Expression of Uncertainty in Measurement' – Propagation of Distributions Using a Monte Carlo Method, JCGM 101:2008, *JCGM*, 2008.

[15]. S. Bouhouche, Z. Mentouri, H. Meradi, Y. Laib, Combined use of support vector regression and Monte Carlo simulation in quality and process control calibration, in *Proceedings of the International Conference on Industrial Engineering and Operations Management (IEOM'12)*, Istanbul, Turkey, 3-6 July 2012, pp. 2156-2165.

[16]. M. Bazil, C. Papadopoulos, D. Sutherland, H. Yeung, Application of probabilistic uncertainty methods (Monte Carlo simulation) in flow measurement uncertainty estimation, in *Proceedings of the International Flow Measurement Conference (IFMC'01)*, 2001, pp. 1-21.

[17]. Y. Fodili, Méthode de contrôle qualité. Application à la fabrication et à la qualification des processus des tubes soudés, *Mémoire d'Ingénieur d'état – Ecole des Mines – Métallurgie Annaba*, 2016.

Chapter 4

Recent Advances in Water Cut Sensing Technology

Prafull Sharma and Hoi Yeung

4.1. Introduction

Multiphase Flow Meters (MPFM) are increasingly being used for both subsea and offshore applications. They are gaining acceptance due to the benefits they bring over test separators, production control and flow assurance to name a few. Multiphase Flow Meters commonly use the measurement of mixture density or electrical permittivity of the produced mixture for estimation of phase fractions [1]. Water cut meter is often part of multiphase flow meters for measurement of water content in the multiphase mixtures. There are standalone water cut meters utilized in the oil and gas production. In this chapter, water cut measurement technologies and methods are discussed and their gaps identified. Microwave based water cut measurement is then described in greater detail along with key academic and industrial research trends.

4.2. Water Cut Measurement Technology

Water cut measurement is an essential component of most multiphase meters and normally is also a part of their final output. By monitoring water cut, operators can optimize the oil production rate, chemicals injection rate, detect water breakthrough among other production and flow assurance operations. Standalone water cut meters are also used in several segments of the oil and gas industry and there are several

---

Prafull Sharma
Chief Technology Officer, Corrosion Radar Ltd, United Kingdom

measurement methods available (Fig. 4.1). Other generic application segments of water cut meters are:

- Upstream: Production well-head, gas-liquid separator, liquid-liquid separator, fiscal and allocation metering, Custody transfer;

- Midstream: Fiscal measurements, condensate water, fiscal metering;

- Downstream: Refinery crude feed, de-salter feed and control, fiscal quantity and quality measurement, shipping terminals, refined products, crude loading and unloading.

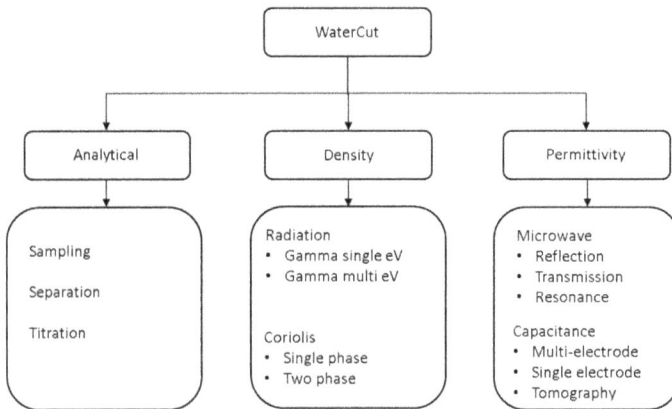

Fig. 4.1. Common principles for water cut measurements.

4.2.1. Analytical Methods

4.2.1.1. Sampling and Centrifugal Separation

In this traditional method, a sample of the oil-water mixture is taken from the production stream through a sampling probe or a sampling port on the pipeline. Then, the oil and water are separated using a centrifuge. Finally, the volume of the separated water is measured, and the water cut of the sample is calculated based on the total volume of the liquid sample.

This method is often used in laboratory tests. It is also widely used to calibrate other water-cut meters. However, the following disadvantages limit its field applications. Firstly, it cannot provide a continuous measurement. In addition, this method is time-consuming, the accuracy is questionable, and personnel are exposed to hazardous chemicals. Note that the sampling, separation and measurement process of live hydrocarbons requires adherence to strict Health, Safety and Environmental (HSE) rules. Most importantly, the water cut measurement accuracy strongly depends upon the integrity of the fluid sample. The sample is affected by many factors such as the phase distribution inside the pipe, location and orientation of the sampling probe and/or sampling port, and the skill of the sampling personnel.

4.2.2. Density Measurement Methods

4.2.2.1. Differential Pressure Method

When an oil-water mixture flows through a vertical pipe with a constant internal diameter, the total pressure drop is composed of two terms: gravitational pressure drop (or hydrostatic pressure drop) and frictional pressure drop. The latter term is often negligible if the mixture velocity is not extremely high. If the total pressure drop through a section of vertical pipe is measured with a differential pressure transducer, the hydrostatic pressure drop of the oil-water mixture can be obtained by ignoring or estimating the frictional pressure drop. The hydrostatic pressure drop of the oil-water mixture through the vertical pipe can be expressed as follows in equation (4.1),

$$\Delta P_g = \mathbf{g}L[\rho_w WC + \rho_o(1 - WC)], \qquad (4.1)$$

where g is the gravitational acceleration, L is the length of the vertical pipe section, WC is the water-cut, and ρ_o and ρ_w are the densities of oil and water, respectively. Clearly, the water-cut can be calculated from equation 4.1 when the oil and water densities are known.

This is a very simple and low-cost method, only requiring a commercially available differential pressure transducer [2]. It gives continuous, real-time measurement with negligible pressure loss. The accuracy of this method is strongly dependent on the difference of oil and water densities. For light oil systems, it gives acceptable measurement accuracy due to the significant density difference between the oil and water phases. The water cut measurement accuracy decreases with a decrease in the difference of oil-water densities. Another disadvantage of this method is that entrained gases will have a detrimental effect on the measurement since a small amount of entrained gases can cause a significant change in the differential pressure across the vertical pipe. An additional problem is in obtaining a good oil density at operating pressure and temperature. Laboratories often report dead oil density at room temperature with no solution gas. This density can be very different from that of flowing live crude oil.

4.2.2.2. Gamma Ray Densitometer

Density-based water cut meters often utilize gamma-ray densitometry which is a nuclear source. However, the adoption of gamma-ray densitometers has been limited for offshore and topside applications due to concerns over the use of a radiation source. The logistics and regulatory procedures around the handling of gamma source make it inconvenient for the end users. Gamma-ray densitometer is a well-adopted sub-component for subsea MPFM and is among the most reliable measurements. Both single energy and dual energy gamma-ray technology are used in the measurement of density and to compensate for additional effects as salinity. Beams of gamma rays are attenuated by the materials through which they pass. The absorption of a beam of initial intensity I_i (photons per square meter per second) is described by an exponential absorption law as follows in equation (4.2)

$$I = I_i \, exp[-\beta z], \tag{4.2}$$

where β is the mass absorption coefficient and z is the distance travelled through a homogeneous absorbing medium.

When applying this technique to an oil-water emulsion flowing inside a pipe, a collimated beam of gamma rays is passed through the pipe wall, through the oil-water-gas three-phase mixture, and through the opposite pipe wall before it reaches the detector. The water-cut of the oil-water-gas emulsion can be determined by the following equation (4.3)

$$I = I_i \, exp\{-(\beta_o\rho_of_o + \beta_w\rho_wf_w + \beta_g\rho_gf_g)d\}, \tag{4.3}$$

where β, ρ and f are respectively the mass absorption coefficient of the gamma-rays, the density and the volume fraction of the mixture component in the gamma-ray path; subscripts o, w and g respectively refer to oil, water and gas; d is the internal diameter of the pipe, and Dual-energy gamma ray absorption technique has been used to measure the water-cut in oil-gas-water three-phase flow. The attenuation of the gamma rays by the fluid is measured at two different energy levels. By performing measurements at two different gamma-ray energies, two equations can be used with different values of mass absorption coefficients. The volume fraction of oil, water and gas can be calculated from these two equations using mass absorption coefficients for the particular gamma-ray energy and knowledge of the densities and composition of the oil, water and gas.

4.2.2.3. Coriolis Densitometer

Coriolis metering technology for water cut and multiphase flow is actively worked upon by several research groups and is continuing to develop rapidly [3]. Coriolis measurement principle is one of the most accurate metering techniques for single phase applications. Since Coriolis meters also measure the density of the fluid passing through it, they can be used in the measurement of water cut in case of oil-water mixtures. A typical Coriolis force flowmeter has two identical tubes that are vibrated in opposition at their natural frequency by an electromagnetic drive mechanism. Because of the Coriolis effect, the fluid flowing through the vibrating tubes creates an asymmetric distortion between the inlet and outlet legs. The distortion magnitude, measured by two position detectors placed on opposite tube legs, is directly proportional to the mass flow rate. Besides mass flow rate, the fluid density can also be determined from the change in vibrating frequency of the meter tubes. A resistance-type temperature sensor continuously monitors the meter tube temperature for various signal processing purposes. Since the Coriolis meter has no moving parts within the flow path, it requires significantly less maintenance.

Water-cut (WC) is calculated from the measured emulsion density (ρ_m) using the definition of mixture density as per the equation (4.4):

$$\rho_m = \rho_w WC + \rho_o(1 - WC), \qquad (4.4)$$

where ρ_o and ρ_w are the densities of oil and produced water, respectively.

Because water-cut is based on density difference between oil and water, measurements can be obtained over the full range of 0-100 % water cut if accurate flowing density is available for the oil and water. This is true regardless of whether the emulsion is oil-continuous or water-continuous.

Although Coriolis flow meters are highly accurate for single phase metering, their performance in oil-water mixtures is not quite trustworthy as shown by several research studies [3]. Moreover, even a small fraction of gas can adversely affect the measurement results.

Since density based water cut measurement is dependent on density contrast between oil and water, they are not a preferred method for heavy oil metering where the density of oil and water are much closer to each other.

4.2.3. Infrared Method

This is a technology based on the bulk transmission of infrared radiation through an oil-water mixture [4]. Its basic principle is spectroscopy which relies on the large difference in absorption of infrared radiation by crude oil and water. Over a very narrow band of radiation, the wavelengths for water are effectively transparent and oil is a strong absorber. The Infrared water-cut meter is a full-range water, and it is unaffected by the transition from oil-continuous phase to water-continuous phase.

4.2.4. Permittivity Measurement Methods

Permittivity based methods (Fig. 4.1) utilize the fact that the permittivity of oil ($\varepsilon_r \sim 2\text{-}3$) is much lower than that of water ($\varepsilon_r \sim 50\text{-}80$). A good review of methods for industrial measurements is provided by [5]. The contrast in permittivity of oil and water is a result of their respective non-polar and polar molecular structures. Since, this contrast in permittivity is maintained irrespective of the type of oil – light or heavy, permittivity based measurements are attractive for a wide range of applications. Moreover, unlike gamma densitometers, Permittivity measurement meters do not use ionizing radiation. This is another factor which makes them attractive for end users especially for topside and offshore applications.

There are primarily two techniques used for permittivity based water cut measurements. RF / Microwave based sensors (100 MHz – 10 GHz) have emerged rapidly for water cut measurements. Yet another popular measurement principle is to utilize capacitance measurements between electrodes to measure fluid permittivity at frequencies below 10 MHz.

4.2.4.1. Capacitance, Conductance & Impedance Principles

Impedance is an important parameter used to characterize electronic circuits, components, and the materials used to make them. The measurement of capacitance can be accomplished by measuring impedance [6]. A typical capacitance probe uses a cell composed of two plates or an open circuited centre rod with a circular pipe as the outside element. Fluid flows through the space separating the inner and outer electrodes [7-10].

The parallel plate capacitance method measures the electrical energy storage based on the dielectric constant of the material between the two plates. The dimensions are fixed and the capacitance value varies only with a change in the insulator fluid (oil) composition which affects the dielectric constant. This can be predicted in terms of the dielectric constant which gives rise to the impedance of the capacitor so formed. A dielectric constant is a dimensionless number; it simply expresses the ability of a material to be polarized and therefore stores electrical energy. Since the measured capacitance (impedance) is proportional to change in dielectric constant (due to water-cut), using the relationship of the dielectric constant to the capacitance, the measured capacitance can be converted to a water content output signal. Analysers based on capacitance measure principles are accurate only when water-cut is relatively low and the oil/water emulsion is oil-continuous (i.e., water droplets in oil-continuous phase). Because these analysers depend on dielectric properties of the emulsion, erroneous measurements occur at higher water cuts even before the emulsion becomes water-continuous (i.e., oil droplets in the water-continuous phase). The relationship is non-linear above approximately 10 % water percentages. In most cases, the highest water-cut of the oil-continuous emulsion occurs at about 35-60 %. If the medium under measurement does not absorb much of the low-frequency energy, capacitance measurements are accurate in the range of water cuts between 0.5 % to 15 %. When the crude oil absorbs energy it is no longer a pure capacitance measurement but, it also has a conductive portion which creates an error in the capacitance measurement. Some crude oils do absorb energy and therefore the measurement becomes more difficult since this further reduces the sensitivity as this becomes a two variable measurement instead of just one variable system. Capacitance measurement is a single factor measurement, it only measures the amount of energy stored between two metal plates, and this limits the capability of this type of measurement.

Imaging techniques have been developed by several researchers using electrical impedance methods. Wiremesh sensors are used in laboratories to study the spatial distribution of phase fractions [11]. Tomographic measurement systems using Electrical Resistance Tomography (ERT), Electrical Capacitance Tomography (ECT) or Electrical Impedance Tomography are getting more matured for field use however they still are largely laboratory measurement techniques [12]. Kvandal *et al.* (2010) and Schuller (2004) [13, 14] described a SECAP sensor which is a non-intrusive single-ended configuration of capacitive water cut measurement. Zhai (2015) [15] described a liquid holdup measurement with double helix capacitance sensor for horizontal oil-water flow.

4.2.4.2. Microwave Principles

Microwaves and Radio Frequency (RF) waves are electromagnetic waves which can travel in dielectric media. Conventionally, the frequency range of 3 MHz – 300 MHz is classified as radio frequency (RF) whereas the spectral range of 300 MHz – 300 GHz is considered to be microwave frequency range. However, in practice, the sensors operating either in RF or microwave are generalized as microwave sensors. Katze and Hubner (2010) [16] reviewed the electromagnetic techniques to water content determination of materials.

Industrial microwave sensors have been in existence for several decades but their development and industrial adoption have accelerated in recent years, mainly due to miniaturization and availability of cheaper components. Many of the new measurement problems which have been tackled by different kinds of microwave sensors are described by Nyfors (2000) [17].

Microwave based water cut sensors fall into the category of material property measurements, where the permittivity of an oil-water mixture is measured to estimate the water content. The complex permittivity and permeability of the medium in which electromagnetic waves propagate affect the wavelength, speed and attenuation of the wave. Gregory and Clarke (2006) [18] described RF and Microwave techniques for dielectric measurements on polar liquids. By measuring the permittivity of a mixture, and using the permittivity of mixture constituents, the phase fraction of constituents can be estimated using an appropriate mixture model.

Microwave sensors are attractive because they are not limited by the health risks associated with radioactive radiation based meters and their fairly low accuracy or the undesirable influence of contamination on the capacitive sensors.

A summary of the water cut measurement methods is show in Table 4.1.

Table 4.1. Summary of the methods for water cut measurement.

Method	Category	Key advantages	Main limitations
Sampling	Analytical	Traditional and robust	Manual, offline and time consuming
Differential Pressure	Density	Differential Pressure measurement is a robust measurement and has large knowledgebase	Accuracy dependent on flow regimes, mixture transition zone and sensitive to gas presence
Gamma Ray Absorption	Density	Robust measurement, Clamp-On configuration	Nuclear radiation concerns users, oil-water density contrast can be low specially for heavy oil
Coriolis Meter	Density	High accuracy and high precision, mature technology for homogenous fluid flow	Sensitive to mixture homogeneity, sensitive to gas presence
Infrared	Permittivity	Spectrometric technique can measure multiple component fractions	Small iris and slot size limits use in heavy oil in poor mixtures
Impedance, Capacitance, Conductance	Permittivity	High dielectric contrast between oil and water	Method range dependent on mixture type such as oil-in-water (o/w) or water-in-oil (w/o)
Microwave	Permittivity	High dielectric contrast between oil and water. Non-ionization radiation is better for health and safety considerations.	Dependency of water cut estimation on mixture models which may be dependent on flow regimes and mixture types

4.3. Microwave Measurement Technology

4.3.1. Industrial Microwave Sensing Principles

The most common water cut measurement types can broadly be categorized based on microwave transmission, reflection and resonance as shown in Fig. 4.2.

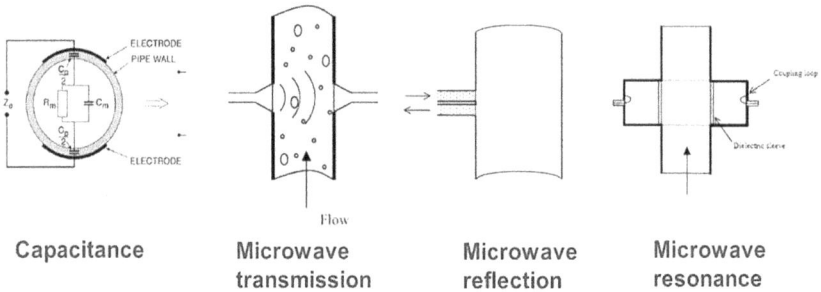

Fig. 4.2. Categories of water cut as described by Nyfors (2000) [20].

There are several possible ways to design and assemble microwave transmitters and receivers in a pipe according to the suitability of end application.

4.3.1.1. Transmission Sensors

In transmission based sensors in a pipe, the transmitter and receiver antenna are placed behind dielectric walls facing towards each other. The electromagnetic wave travels between the antennas while penetrating through the flowing media in the pipe. The complex permittivity of the media affects both the phase and the amplitude of the travelling wave [19]. The advantage of this arrangement is the simplicity and good sensitivity. But the main challenge is the complications in signal interpretation which is also influenced by reflections in various parts of the system, like the dielectric walls and other interfaces arising out of flow regimes. The reflections in the system may lead to difficulties in signal interpretation and can cause inaccuracy in estimation [20].

4.3.1.2. Reflection Sensors

In reflection based sensor system, amplitude and phase of the reflected wave from the end of the transmission line are measured [21]. The measurement principle is commonly based on the influence of flowing media on the fringing electromagnetic fields which are in direct contact with it. A common example is the open-ended coaxial probe, which is used for measuring permittivity over a broad frequency range [22, 23]. A disadvantage of reflection type sensors is the small sensing volume. Due to this, it is sensitive to drift due to deposits over it. Because of the small local sensed volume, it is not a suitable principle for inhomogeneous mixtures [24].

4.3.1.3. Resonator Sensors

The resonance frequency of a microwave resonator is related to the permittivity of the media in the resonator. By measuring the resonance frequency, the permittivity of the flowing media can be estimated [25, 26]. Resonance based measurements are quite sensitive to even small changes in permittivity, however, this can also be a disadvantage when the wide range is to be measured [27]. One of the distinct advantages of the resonator based sensor is that it is relatively robust as compared to amplitude and phase measurements. Because microwave resonators are inherently stable and the resonant frequency and quality factor (Q-factor), which are the two measurable properties of a resonance, can be measured with a high accuracy, the microwave resonance method is the most sensitive and accurate method available for measuring the Water Volume Fraction (WVF) of a wet gas flow [5, 25].

Microwave resonators can be arranged in several different ways inside a pipe. There are broadly two classes of resonator sensors a) strongly coupled to media b) weakly coupled to media. The former class is more suitable for materials with low losses. For water cut measurements, this generally means oil-continuous fluids. In case of highly lossy media, significant damping can result in the elimination of resonance. The resonators with a weak coupling to the media work better in such a case [25, 28]. In weakly coupled resonators, only a small fraction of the resonance fields are exposed to the media resulting in just a perturbation to the resonance resulting in a reduced sensitivity of the sensor. Nyfors

157

(2013) [47] described an arrangement where a dielectric resonator was used to sense just near the wall of the pipe.

4.3.2. Water Cut Estimation from Microwave Sensors

The Microwave based sensing principles often use a water cut estimation process as shown in Fig. 4.3. The other way is a direct calibration model from measurement to water cut developed in calibration facilities and applied in the field. But that approach suffers from bias when fluid properties of a field are different from that of calibration facility. Hence a model-based approach is better (Fig. 4.3). The measured microwave sensing parameters from the sensor at a single or multiple frequencies are used to estimate the media permittivity of the oil-water liquid mixture. The water permittivity (ε_w) and oil permittivity (ε_o) along with oil-water mixture permittivity (ε_m) are used in a mixture model to calculate water cut. Water permittivity can be calculated using the model developed by various researchers [29, 30] for known water temperature, salinity and the electrical frequency. There are several mixture models that can be used to relate permittivity of oil, water, the oil-water mixture to Water-Liquid Ratio (WLR) [31].

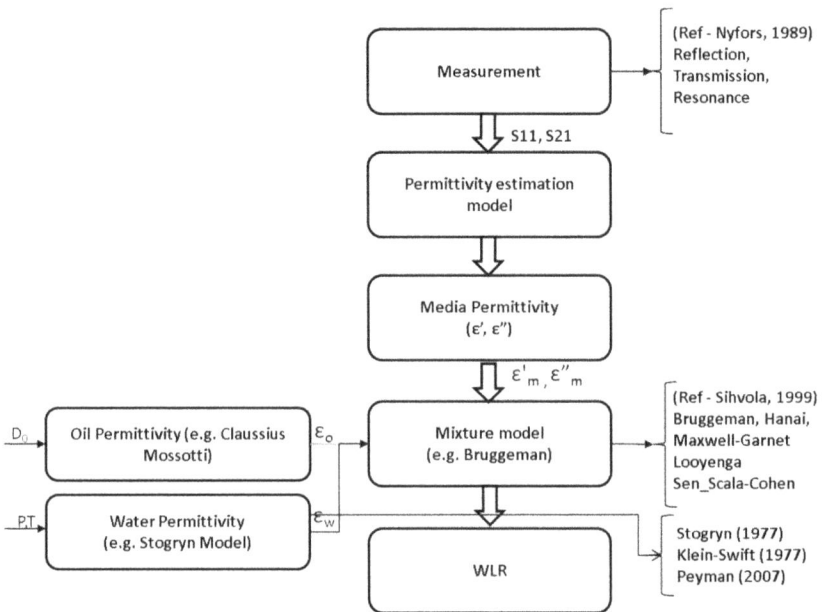

Fig. 4.3. Estimation process of water cut from microwave measurements.

4.3.3. Advances in Microwave Sensing Technology

One of the earliest developments on Radio Frequency based sensor was made by Dykesteen *et al.* (1985) [32] consisted of a method of non-intrusive measurement of fluid fractions using two insulated electrodes and flowing multiphase mixture between them. The electrical impedance is measured and used to estimate the fractions of water, oil and gas. Methods based on electrical capacitance have also been developed, though measurement uncertainty and instrument drift issues limit their usage for high accuracy requirements. Hammer *et al.* (1989) [10] demonstrated a helically shaped capacitance electrode system for water fraction in oil-water mixtures with repeatability better than 1.5 %. They also demonstrated that helically shaped electrodes are less dependent on variation in flow regimes. Demori *et al.* (2010) [33] proposed a solution to the problem of capacitive sensing in the presence of conductive water, which introduces parasitic coupling to stray elements outside the measurement section of the pipe. They proposed a novel sensor configuration that employs guard electrodes, coupled to a tailored electronic interface to drive the guard electrodes.

Electrical tomography based methods have been gaining interest especially for multiphase flow imaging and phase fraction estimation, although industrial versions for field use are still being developed. Yang (2010) [34] extensively reviewed the electrical capacitance tomography technology and provided design guidance. The measurement problem is particularly complex at the lower end of the WLR range (0-5 %), where poor contrast in dielectric permittivity (ε_r) between the oil-water mixture and the oil alone proves to be a challenge for applications demanding high accuracy [35].

Thanks to the increasing availability of precise electronics, driven by developments in the telecom industry, microwave-based sensing solutions for robust and accurate WLR measurement have been emerging rapidly. Castle *et al.* (1974) developed one of the earliest microwave-based sensing systems for water fraction measurement in crude oil applications. Krupa (2006) [36] provided a comprehensive review of sensing methods for microwave in frequency domain.

Zarifi *et al.* (2016) [37] established a microwave planar ring resonator sensor tuned at 5.25 GHz, providing a non-contact method for liquid-liquid interface detection. They demonstrated the applicability of this sensor to interface detection in water-olive oil-ethanol samples representing a wide range of permittivity. The resonator's Quality Factor (Q-Factor) is also used in the estimation process, in addition to the resonance frequency, to increase the robustness of the sensor.

Oon *et al.* (2016) [28] investigated a cylindrical microwave resonator sensor to monitor two phase flow systems and the changes in the permittivity of the measured phases to differentiate between the volume fractions of air, water and oil. Microwaves in the range of 5-5.7 GHz have been used to analyse a two-phase air-water and oil-water stratified flow in a pipeline, demonstrating the ability to detect a change in water fraction in full range of 0 % – 100 %.

Zarifi and Daneshmand (2016) [38] proposed a non-contact liquid sensor using an active, feedback loop assisted, planar, micro-strip microwave resonator. The sensor has the ability to operate in a non-contact fashion within a distance of 0 to 8 cm. The active loop technique is shown to increase the primary Q-Factor from 210 to 500,000 in air when measured at a resonant frequency of 1.52 GHz. The proposed device is used to distinguish between water, ethanol, methanol, isopropanol, and acetone in a submerged tube inside a water-filled container. They also demonstrated the application of micro-strip microwave resonator sensor in monitoring solid particle deposition in lossy medium [39].

Al-Kizwini *et al.* (2013) [40] proposed a non-intrusive sensor, which is based on an electromagnetic waves cavity resonator. It determines and monitors the percentage volumes of each of the two phases (oil and gas) in the pipeline using the resonant frequencies shifts that occur within the resonator. Temperature has a significant influence on the liquid permittivity, especially for water. It also affects the resonance frequency of the measurement by microwave resonators and hence it is important to compensate the measurements against variations due to temperature [41].

Surface perturbation methods have gained research attention due the near wall measurement they provide that can be effective in measuring liquid properties even in the presence of gas. Furthermore, resonance based microwave sensors are evaluated for demanding applications, such as waste water with very low concentration of oil [42]. Ni and Ni (1997) [43] described a class of non-intrusive microwave resonator sensors, called extra-cavity perturbation methods, for generic applications.

Jannier *et al.* (2013) [44] presented a microwave reflection method to measure the water content in oil. Microwave reflectometry is applied to multiphase flow metering in the context of oil extraction. The sensor consists of two open-ended coaxial probes operating at complementary frequencies (at 600 MHz and around 36 GHz) and was designed to resist harsh field conditions. This publication presents and comments on results obtained in realistic dynamic conditions, on a three-phase flow loop (water-oil-gas). The main conclusions are the following: Bruggeman and Hanai's mixing rule applies to natural emulsions and can be used to determine the composition of the water-oil liquid phase; results obtained for annular flows are very sensitive to small perturbations such as bubbles or waves at the liquid-gas interface; in the case of triphasic slug flows, the composition of the liquid phase can be estimated by proper filtering of the data.

Tan *et al.* (2015) [45] described an information fusion approach to measure multiphase flow parameters including water cut. They used a conductance ring sensor and combined it with pressure drop measurement for estimation. This work indicates a research trend of using multiple sensor data fusion to estimate the parameters of interest. Zhang *et al.* (2014) [46] also described a soft measurement estimation of water holdup based on neural network models.

In the patent by Nyfors (2013) [47], an open-ended cylindrical resonator is described for near wall measurement of water salinity in wetgas in a multiphase stream. In a wet gas flow, it is important to know the WVF because of the problems with hydrate formation, scaling, and corrosion caused by the water. Also, the salinity of the water, caused by the production of formation water, is a very important factor as it strongly affects both corrosion and the formation of scale.

The microwave resonance principle is based on measuring the permittivity / dielectric constant of the flow. Water, being a polar molecule, have a high permittivity (40-80) compared to that of oil (in the order of 2-3) or gas (~1) the permittivity of a mixture of these three constituents is dominated by the contribution from the water. Methods based on measuring the permittivity (microwave and capacitive methods) therefore provide the highest sensitivity for measuring the WVF of a mixture. Because microwave resonators are inherently stable and the resonant frequency and quality factor (Q-factor), which are the two measurable properties of a resonance, can be measured with a high accuracy, the microwave resonance method is the most sensitive and accurate method available for measuring the WVF of a wet gas flow. However, when the WVF becomes very low, the permittivity of the mixture (i.e. the flowing fluid) starts to become dominated by the contributions from the oil and the gas. Especially the permittivity of the gas depends on the pressure and the temperature. To be able to resolve the contribution from the water one needs to know the contributions from the gas and the oil. For example, the invention described by Nyfors and Oystein (2005) [48] also uses the hydrocarbon composition and measurements of temperature and pressure as inputs, and models for calculating the permittivity of the oil and the gas. The accuracy of the measurement of the WVF is then limited by the accuracy of the models, and the accuracy of the measurements of the temperature and pressure.

Wylie *et al.* (2006) [49] developed an intrusive electromagnetic cavity resonator based sensor for multiphase flow measurement through an oil pipeline. This sensor is non-intrusive and transmits low power (10 mW) radio frequencies (RF) in the range of 100-350 MHz and detects the pipeline contents using resonant peaks captured instantaneously. The multiple resonances from each captured RF spectrum are analysed to determine the phase fractions in the pipeline.

Nyfors (2000) [5] developed a downhole water cut meter. This sensor is based on semi-sectorial resonance modes and is designed to measure the mixture ratio of oil and water in an oil well deep in the ground where the temperature and pressure are high. Because the space between the casing and production tubing is annular, the shape of the sensor was designed to match the spatial requirements in the annulus of an oil well. This resonance measurement method is not accurate when the sample is a high-loss material.

An example of a method for measuring properties of flowing fluids and a metering device and a sensor used for performing this method has been described in [5]. The sensor uses the microwave resonance principle for the measurement of oil-continuous fluids (water drops and gas bubbles in oil, i.e. the oil is a continuous phase) and the measurement of conductivity for water-continuous fluids (oil drops and gas bubbles in water, i.e. the water is the continuous phase, and is intended for installation in a production zone inside an oil well.

Another example of a method for measuring flowing fluids with a far higher gas content, i.e. wet gas (a wet gas flow is a multiphase flow with a high gas volume fraction, usually called the gas void fraction (GVF), typically > 99 %) or high-gas multiphase flow, has been described in patent by [47]. This is also based on the microwave resonance so principle.

Huang (1998) [50] developed a transmission line model based on microwave theory to improve the calibration procedure of the water cut meter, Starcut. Through this microwave model and mixing law, the attenuation and phase shift of microwave signal can be directly related to water cut of oil-water mixture. This model uses empirical formulae to consider hole effect problem and did not include microwave energy leakage along sample pipe. This will result in inaccuracy of results from model prediction.

Folgero and Tjomsland (1996) [51] proposed open-ended coaxial probe to measure the composition of a well-mixed water-oil emulsion layer. The composition of this thin emulsion layer can be determined from permittivity measurements with this open-ended coaxial cable. Techniques for probe design and determination of liquid permittivity are also studied. They developed a bilinear model and an empirical exponential model to estimate the permittivity of thin liquid layers. This bilinear model needs calibration measurements on three samples with known permittivity. A full-wave model of the probe is also developed with lower precision, but this model need not be calibrated. A simple reference plane rotation is needed. This method also cannot be used to study characteristics of interior section fluids such as oil and water droplet shape, interface property and emulsion status.

Marrelli *et al.* (1996) [52] developed a microwave water cut monitor for field application and for the continuous measurement of the percent water in crude production streams over the entire 0 to 100 % range. This monitor can continuously report emulsion status (water-continuous or oil-continuous), effective NaCl salinity and effective paraffin carbon number of the hydrocarbon phase. Operator interaction is minimized by features such as automatic self-calibration, unattended operation at remote locations, remote access options through SCADA systems and self-starting after power outages. However, the calibration method of this monitor is to compare the data from the unknown sample with previously measured data which came from a known sample and were saved in the calibration database. Because of the complexity of multiphase flow, this calibration process may affect measurement results. Existing water cut predominantly depends on empirical data and correlations that are sensitive to fluid properties. Correlation methods are therefore limited in their general applicability and require frequent re-calibration.

4.3.4. Research Challenges in Microwave Water Cut Metering

In recent years, significant technology advancement has been done for both MPFM and standalone water cut measurement. Several manufacturers have been continuously testing and maturing the technology and new suppliers are also rapidly coming up with novel solutions. However, it is a general consensus among industry experts that a universal meter is a challenge. A lot of improvement is still required to achieve accurate, reliable and robust MPFM and water cut metering. Below are a few known gaps in metering technology

4.3.4.1. Flow Regimes

The performance of MPFM and water cut meter is highly dependent on flow regimes they operate in. As the gas fraction increases above 90 %, the accuracy of the MPFM drops significantly. There are separate wetgas meters available which are especially suited to perform measurements at high GVF conditions (>95 %). The water cut measurement in high GVF is challenging due to a small liquid content in the flow [53, 54].

4.3.4.2. Periodic Calibration

MPFM and water cut meters suffer from changes in fluid compositions. Due to this, the fluid properties changes over a period of time resulting in the need for recalibration of the sensors periodically. Periodic calibration poses a logistic and operational constraint on operators [55, 56].

4.3.4.3. Oil Composition

The changes in oil composition over a period of time changes its physical and electrical properties resulting in deviations from values originally fed into the meter. Hence, the change in properties as viscosity, density and permittivity affect the accuracy of a meter [55-57].

4.3.4.4. Scaling and Fouling

Over time, scale can develop inside the meter spool surface which changes the sensor performance especially if a reflection type sensor is used in the meter. If a DP-Venturi is used, the coefficient of discharge can change due to scale [53]. Moreover, due to scaling the effective cross-section area available to the flow changes which the does not account for. These factors may lead to deviation in the accuracy of the meter. Wee *et al.* (2013) [58] demonstrated scale formation inside a meter after several years of field operation.

4.3.4.5. Dielectric Mixture Models

Dielectric mixture models can cause significant errors in relating measured signals to permittivity. These models often assume that the mixture is made up of simple geometrical inclusions like spheres, ellipsoids of droplet phase within a host (continuous) phase as described by Sihvola (1999) [31]. The non-linear behaviour of various mixture models are as shown in Fig. 4.4. They also often assume a homogenous distribution of the droplets in the host media. In the real field conditions, there may be significant deviations from the simplistic assumptions made by dielectric mixture models. This can be a potential cause of inaccuracy depending on the application. This is especially true in case of poorly mixed flows and is one of the reasons of installation of mixers before the meters.

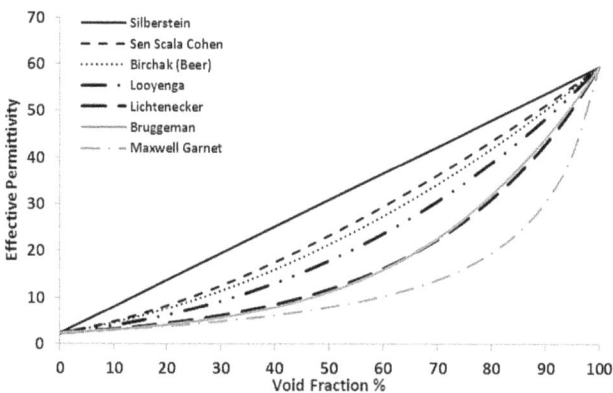

Fig. 4.4. Dielectric mixture models for a mixture of two dielectric materials. Void fraction is of the particle phase dielectric component (Shivola, 1999) [31].

4.3.4.6. Emulsion Transition Zone

The transition point of the emulsion type from water-in-oil (w/o) to oil-in-water (o/w) or vice versa is not well understood and is extremely hard to predict (Figs. 4.5, 4.6). This zone poses a big measurement challenge as the mixture properties are not well understood leading to poor performance of the meters [54].

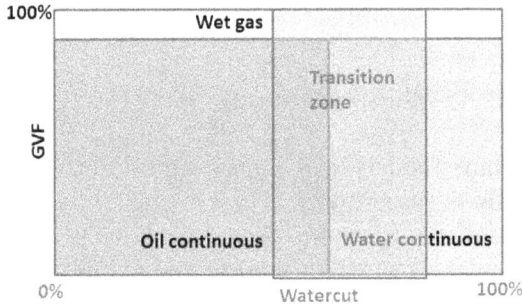

Fig. 4.5. Transition zone of the oil-water mixture is a measurement challenge.

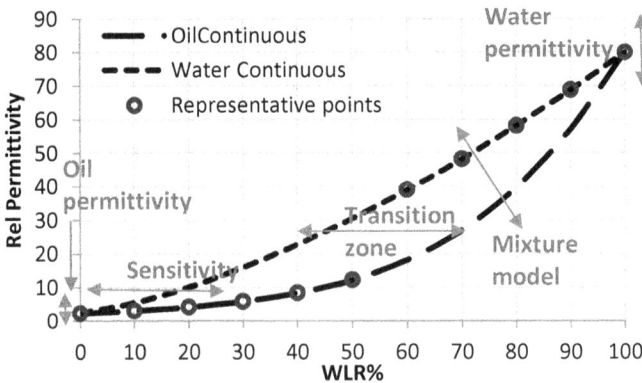

Fig. 4.6. A pictorial representation of a few factors causing water cut errors.

4.3.4.7. Salinity

Presence of salt alters the permittivity (Fig. 4.6) and density of aqueous phase and hence affects the estimation of water cut. In oil-in-water flow,

MPFMs need the input of water conductivity/salinity values in order to perform as per their specifications. Water-in-oil mixture has the much lesser influence of salinity change as compared to that of oil-in-water [53]. With a change in salinity over time, the meter readings tend to deviate from their correct values. Due to this, several research groups and manufacturers have been developing their own proprietary salinity sensors [17, 59, 60]. In case of microwave sensors, microwave transmission across a pipe may be significantly attenuated to measure. Hence, highly saline water in oil-in-water mixture poses a challenge for transmission based microwave sensors [61, 62]. Seraj *et al.*(2014) [60] reviewed the state-of-the-art in sensor technologies to measure salinity in multiphase production streams. Saetre *et al.*(2010) [63] demonstrated a dual-energy gamma based measurement system for salinity independent measurement. Somaraju and Trumpf (2006) [64] developed a model to estimate water dielectric properties with various salinities.

4.3.4.8. Hydrate Inhibitors

To prevent hydrate formation, a significant amount of thermodynamic hydrate inhibitors namely Methanol and Mono-Ethylene Glycol are injected at the well head. Since these are miscible in water, they are measured as water by the microwave as well as gamma-ray densitometers. These chemicals significantly affect the permittivity and density of water thereby affecting readings from the meters [55, 65]. Recent developments using Infrared Absorption have claimed to measure hydrate inhibitors in multiphase mixtures as well as independently estimate water cut [4].

4.3.4.9. Sensitivity at Low Water Cut

The sensitivity of water cut with respect to the permittivity of the oil-water emulsion at low water cut is quite low as compared to sensitivity at higher water cut (Fig. 4.6). This is challenging as the sensors should be designed to handle a wide dynamic range.

4.3.4.10. Heavy Oil

Heavy oil fields are has been increasingly adopting the use of multiphase flow meter but the performance and accuracy of meters are still not in desirable limits [66]. Due to the high viscosity of heavy oil, the flow through the meter is more likely to be in the laminar region rather than turbulent. This may be difficult to validate in the field conditions and hence some of the key assumptions in meters may not hold true [55]. Moreover, the emulsion characteristics of heavy oil at field conditions are not well understood. Pinguet (2011) [57] described the challenges with Heavy oil metering and reviewed several measurement technologies.

4.3.4.11. Entrained Gas

A common problem encountered in oil well production testing is gas "carry-under" into the separated liquid stream. This results in a trace amount of gas entering the water-cut meter, producing errors in the water-cut reading [44]. Gas carry-under may be caused by high liquid viscosity, improper separator operation, a separator with a build-up of sand in the bottom, or poor separator design. These are some of the issues which significantly affect the allocation factors encountered in oil and gas operations. Multiphase measurement is typically performed on the surface, where pressures are relatively low. As pressures are reduced, gas dissolved in the oil is released. A reduction of pressure may cause gas to flash out of solution as crude oil flows from the separator. The amount of gas evolved depends on crude properties, operating temperature, pressure, and differences in pressure across the multiphase. Once entrained gas enters a water-cut meter, it will affect the reading regardless of the principle of operation. When a mass flow meter is used for water-cut measurement, the entrained gas will cause a decrease in density reading, which will be misinterpreted as a decrease in water percentage. The microwave water-cut meter, which operates based on the difference in dielectric constant of water (40 to 80), oil (2-3) and gas (1), will see a decrease in the dielectric constant of the flow stream when a small amount of gas is entrained.

4.3.4.12. Other Factors

There are several other chemicals which are injected at the well head as part of flow assurance strategies. These include emulsifiers, corrosion inhibitors, H_2S, CO_2 among others. Sand erosion can lead to erosion to the throat of venturi [55].

Table 4.2 shows the summary of the research gaps discussed.

Table 4.2. Summary table of water cut metering research gap.

Factors	Issues	Research Gaps	References
Flow Regimes	Flow regimes affect the accuracy of measurements	Effect of flow regimes on transducers	[53, 55]
Periodic Calibration	Drift in sensor demands periodic calibration	Field calibration, self-calibration	[56]
Oil Composition	Field variations of oil properties (viscosity, density, permittivity etc.)	Effect and mitigation on measurement techniques	[57]
Scaling and Fouling	Scaling and fouling over sensors affects accuracy in measurements	Effect and mitigation on measurement techniques	[58]
Mixture Models	Multiphase mixtures are often in-homogenous dielectric mixtures, may violate assumptions behind mixture models	Realistic mixture models derived for in-homogenous mixtures and larger inclusions	[31]
Transition Zone	Transition between mixture types affects accuracy of measurements	Insufficient understanding of transition zone between mixture types	[54]
Salinity	Field variations of water salinity affects meters accuracy	Methods to estimate salinity in multiphase flow in various mixture types	[53, 59, 60]
Hydrate Inhibitors	Flow assurance chemicals affects fluids properties leading to meter accuracy	Effect of Hydrate Inhibitors on multiphase fluid properties and compensation methods	[4]
Sensitivity	Non-linear mixture models shows poor sensitivity at lower water cut	Sensing methods capable of handling wide sensitivity range	[31]
Heavy Oil	Viscosity, mixture properties, flow characteristics in heavy oil affects meter accuracy	Insufficient understanding of heavy oil flow characteristics in field conditions	[65]
Entrained Gas	Entrained gas alters the liquid properties	Factors affecting entrained gas in field and separation challenges	[44]
Other Factors	Injected chemicals affects fluids properties, erosion, corrosion, sand, H_2S	Field factors altering meter characteristics in unpredictable ways	[55]

4.4. Summary

Water cut measurement technologies and methods are discussed and their gaps identified. Microwave based water cut measurement are gaining popularity in academic and industrial communities. Though the research trends indicate that microwave sensing technology is gaining attention and gaining maturity, there are several issues yet to be addressed for accuracy and robustness in field use. Microwave sensors are simplistically categorised into one of the following categories – transmission, reflection, resonance. However, a combination of reflection and resonance is explored only for limited applications in multiphase flows (wet gas, salinity) and the literature for applications in water cut measurements is very limited. There is a quest for microwave water cut sensing methods which are not affected or minimally affected by the presence of gas in the multiphase flow stream. Moreover, the sensing method should also address the low sensitivity of mixture permittivity at the lower end of water cut (also termed as a WLR in this work) and should ideally work in water continuous (oil-in-water) mixtures.

References

[1]. G. Falcone, G. Hewitt, C. Alimonti, Multiphase Flow Metering: Principles and Applications, Vol. 54, *Elsevier*, 2009.

[2]. Y. F. Han, *et al.*, Differential pressure method for measuring water holdup of oil-water two-phase flow with low velocity and high water-cut, *Experimental Thermal and Fluid Science*, Vol. 72, 2016, pp. 197-209.

[3]. M. Henry, *et al.*, New applications for Coriolis meter-based multiphase flow metering in the oil and gas industries, in *Proceedings of the 10th International Symposium of Measurement Technology and Intelligent Instruments (ISMTII'11)*, 2011, pp. 1-6.

[4]. J. Lievois, V. Ramakrishnan, B. Adejuyigbe, Measuring relative concentration of water, condensate and methanol/glycol in a wet gas stream using infrared absorption, in *Proceedings of the Americas Workshop*, 2010, pp. 1-17.

[5]. E. G. Nyfors, Method for Measuring Properties of Flowing Fluids, and a Metering Device and a Sensor Used for Performing This Method, USPTO, Patent US6826964, *USA*, 2004.

[6]. J. Cho, M. Perlin, S. L. Ceccio, Measurement of near-wall stratified bubbly flows using electrical impedance, *Measurement Science and Technology*, Vol. 16, Issue 4, 2005, pp. 1021-1029.

[7]. N. Libert, R. E. M. Morales, M. J. da Silva, Capacitive measuring system for two-phase flow monitoring. Part 2: Simulation-based calibration, *Flow Measurement and Instrumentation*, Vol. 50, 2016, pp. 102-111.

[8]. Z. An, *et al.*, Liquid holdup measurement in horizontal oil-water two-phase flow by using concave capacitance sensor, *Measurement: Journal of the International Measurement Confederation*, Vol. 49, Issue 1, 2014, pp. 153-163.

[9]. M. Z. Aslam, T. T. Boon, Differential capacitive sensor based interface circuit design for accurate measurement of water content in crude oil, in *Proceedings of the 5th International Conference on Intelligent and Advanced Systems: Technological Convergence for Sustainable Future (ICIAS'14)*, 2014, pp. 2-7.

[10]. E. A. Hammer, J. Tollefsen, K. Olsvik, Capacitance transducers for non-intrusive measurement of water in crude oil, *Flow Measurement and Instrumentation*, Vol. 1, Issue 1, 1989, pp.51-58.

[11]. M. J. Da Silva, *et al.*, Phase fraction distribution measurement of oil-water flow using a capacitance wire-mesh sensor, *Measurement Science and Technology*, Vol. 22, Issue 10, 2011, 104020.

[12]. Y. Faraj, *et al.*, Measurement of vertical oil-in-water two-phase flow using the dual-modality ERT-EMF system, *Flow Measurement and Instrumentation*, Vol. 46, 2015, pp. 255-261.

[13]. H. K. Kvandal, S. A. Kjolberg, R. B. Schiller, Water detection in gas/condensate flows by SeCaP technology, *Chemical Engineering Journal*, Vol. 158, Issue 1, 2010, pp. 19-24.

[14]. R. B. Schuller, *et al.*, Measurement of water concentration in oil/water dispersions with a circular single-electrode capacitance probe, *IEEE Transactions on Instrumentation and Measurement*, Vol. 53, Issue 5, 2004, pp. 1378-1383.

[15]. L. Zhai, *et al.*, Liquid holdup measurement with double helix capacitance sensor in horizontal oil-water two-phase flow pipes, *Chinese Journal of Chemical Engineering*, Vol. 23, Issue 1, 2015, pp. 268-275.

[16]. U. Kaatze, C. Hübner, Electromagnetic techniques for moisture content determination of materials, *Measurement Science and Technology*, Vol. 21, Issue 8, 2010, 082001.

[17]. E. G. Nyfors, Cylindrical microwave resonator sensors for measuring materials under flow, DST Thesis, *Helsinki University of Technology*, 2000.

[18]. A. P. Gregory, R. N. Clarke, A review of RF and microwave techniques for dielectric measurements on polar liquids, *IEEE Transactions on Dielectrics and Electrical Insulation*, Vol. 13, Issue 4, 2006, pp. 727-743.

[19]. Y. V. Makeev, A. P. Lifanov, A. S. Sovlukov, Microwave measurement of water content in flowing crude oil with improved accuracy, in *Proceedings of the 24th International Crimean Conference Microwave and Telecommunication Technology (CriMiCo'14)*, 2014, pp. 956-957.

[20]. E. Nyfors, Industrial Microwave Sensors – A Review, *Subsurface Sensing Technologies and Applications*, Vol. 1, Issue 1, 2000, pp. 23-43.

[21]. E. Bondet, *et al.*, Low (10-800 MHz) and high (40 GHz) frequency probes applied to petroleum multiphase flow characterization, *Measurement Science and Technology*, Vol. 19, Issue 5, 2008, 055602.

[22]. K. Folgerø, T. Tjomsland, Permittivity measurement of thin liquid layers using open-ended coaxial probes, *Measurement Science and Technology*, Vol. 7, Issue 8, 1996, pp. 1164-1173.

[23]. T. Jakobsen, K. Folgerø, Dielectric measurements of gas hydrate formation in water-in-oil emulsions using open-ended coaxial probes, *Measurement Science and Technology*, Vol. 8, Issue 9, 1997, pp. 1006-1015.

[24]. X. Chen, *et al.*, Water holdup measurement of oil-water two-phase flow with low velocity using a coaxial capacitance sensor, *Experimental Thermal and Fluid Science*, Vol. 81, 2017, pp. 244-255.

[25]. S. Al-Hajeri, *et al.*, Real time EM waves monitoring system for oil industry three phase flow measurement, *Journal of Physics: Conference Series*, Vol. 178, Issue 1, 2009, 012030.

[26]. B. V. Lunkin, V. I. Mishenin, N. A. Kriksunova, Determination of the volume content of the components of an oil-water type emulsion from the parameters of an electromagnetic resonator, *Measurement Techniques*, Vol. 50, Issue 9, 2007, pp. 1005-1013.

[27]. J. A. Cuenca, Characterisation of powders using microwave cavity perturbation, PhD Thesis, *Cardiff University*, 2015.

[28]. C. S. Oon, *et al.*, Experimental study on a feasibility of using electromagnetic wave cylindrical cavity sensor to monitor the percentage of water fraction in a two phase system, *Sensors and Actuators A: Physical*, Vol. 245, 2016, pp. 140-149.

[29]. L. Klein, C. Swift, An improved model for the dielectric constant of sea water at microwave frequencies, *IEEE Journal of* Ocean Engineering, Vol. 2, Issue 1, 1977, pp. 104-111

[30]. A. Peyman, C. Gabriel, E. H. Grant, Complex permittivity of sodium chloride solutions at microwave frequencies, *Bioelectromagnetics*, Vol. 28, Issue 4, 2007, pp. 264-274.

[31]. A. Sihvola, Electromagnetic Mixing Formulas and Applications, *Institution of Engineering and Technology*, 1999.

[32]. E. Dykesteen, A. Hallanger, E. A. Hammer, E. Samnøy, R. Thorn, Non-intrusive three-component ratio measurement using an impedance sensor, *Journal of Physics E: Scientific Instrumentation*, Vol. 18, 1985, pp. 540-544.

[33]. M. Demori, V. Ferrari, D. Strazza, P. Poesio, A capacitive sensor system for the analysis of two-phase flows of oil and conductive water, *Sens. Actuators A: Phys.*, Vol. 163, Issue 1, 2010, pp. 172-179.

[34]. W. Q. Yang, Design of electrical capacitance tomography sensors, *Meas. Sci. Technol.*, Vol. 21, Issue 4, 2010, 042001.

[35]. I. Ismail, J. C. Gamio, S. F. A. Bukhari, W. Q. Yang, Tomography for multi-phase flow measurement in the oil industry, *Flow Measurement and Instrumentation,* Vol. 16, Issue 2, 2005, pp. 145-155.

[36]. J. Krupka, Frequency domain complex permittivity measurements at microwave frequencies, *Measurement Science and Technology*, Vol. 17, Issue 6, 2006, pp. R55-R70.

[37]. M. H. Zarifi, M. Rahimi, M. Daneshmand, T. Thundat, Microwave ring resonator-based non-contact interface sensor for oil sands applications, *Sensors and Actuators B: Chemical*, Vol. 224, pp. 632-639.

[38]. M. H. Zarifi, M. Daneshmand, Liquid sensing in aquatic environment using high quality planar microwave resonator, *Sensors and Actuators B: Chemical*, Vol. 225, 2016, pp. 517-521.

[39]. M. H. Zarifi, M. Daneshmand, Monitoring solid particle deposition in lossy medium using planar resonator sensor, *IEEE Sensors Journal,* Vol. 17, Issue 23, 2017, pp. 7981-7989.

[40]. M. A. Al-Kizwini, *et al.*, The monitoring of the two phase flow-annular flow type regime using microwave sensor technique, *Measurement*, Vol. 46, 2013, pp. 45-51

[41]. S. Almuradi, *et al.*, Temperature impact in electromagnetic non-invasive water/oil/gas multiphase real time monitoring, *Asian Journal of Eng. and Tech.*, Vol. 3, Issue 5, 2015, pp. 512-527.

[42]. O. Korostynska, A. Mason, A. Al-Ahrama, Microwave sensors for the non-invasive monitoring of industrial and medical applications, *Sensor Review*, Vol. 34, Issue 2, 2014, pp. 182-191.

[43]. E. H. Ni, Y. Q. Ni, Extra-cavity perturbation method for the measurement of dielectric resonator materials, *Review of Scientific Instruments*, Vol. 68, Issue 6, 1997, pp. 2524-2528.

[44]. B. Jannier, O. Dubrunfaut, F. Ossart, Application of microwave reflectometry to biphasic flow characterization, *Measurement Science and Technology*, Vol. 24, 2013, 025304.

[45]. C. Tan, *et al.*, A Kalman estimation based oil-water two-phase flow measurement with CRCC, *International Journal of Multiphase Flow*, Vol. 72, 2015, pp. 306-317.

[46]. D. Zhang, B. Xia, Soft measurement of water content in oil-water two-phase flow based on RS-SVM classifier and GA-NN predictor, *Measurement Science Review*, Vol. 14, Issue 4, 2014, pp. 219-226.

[47]. E. G. Nyfors, Flow Measurements, USPTO, Patent US8570050B2, *USA*, 2013.

[48]. E. G. Nyfors, O. Bo Lund, Compact Flow Meter, USPTO, Patent US6915707B2, *USA*, 2005.

[49]. S. R. Wylie, A. Shaw, A. I. Al-Shamma'a, RF sensor for multiphase flow measurement through an oil pipeline, *Measurement Science and Technology*, Vol. 17, Issue 8, 2006, pp. 2141-2149.

[50]. S. Huang, Computer simulation of microwave flow meter, MS Thesis, *University of Houston*, 1998.

[51]. K. Folgerø, A. L. Tomren, S. Froyen, Permittivity calculator: Method and tool for calculating the permittivity of oils from PVT data, in *Proceedings of the 30th North Sea Flow Measurement Workshop*, 2012.

[52]. J. D. Marelli, Three phase metering in Sumatra using the StarCut meter – An example of a perfect fit to a specialized niche, in *Proceedings of the 14th North Sea Flow Measurement Workshop*, 1996.

[53]. W. X. Liu, *et al.*, Effects of flow patterns and salinity on water holdup measurement of oil-water two-phase flow using a conductance method, *Measurement: Journal of the International Measurement Confederation*, Vol. 93, 2016, pp. 503-514.

[54]. G. Falcone, G. Hewitt, C. Alimonti, Multiphase Flow Metering. Principles and Applications, Vol. 54, *Elsevier*, 2009.

[55]. C. Letton, L. H. Group, A. Hall, Multiphase and wet gas flow measurement – It' s not that simple, *Proceedings of Society of Petroleum Engineers*, 2012, pp. 11-14.

[56]. E. Dahl, *et al.*, Handbook of Water Fraction Metering, *Norwegian Society for Oil and Gas Measurement (NFOGM)*, 2004.

[57]. B. G. Pinguet, Worldwide review of 10 years of the multiphase meter performance based on a combined nucleonic fraction meter and venturi in heavy oil, in *Proceedings of the 29th International North Sea Flow Measurement Workshop*, 2011.

[58]. A. Wee, Ø. Fosså, V. R. Midttveit, A multiphase meter capable of detecting scale on the pipe wall and correcting flow rate measurements, in *Proceedings of the 31st North Sea Flow Measurement Workshop*. 2013, pp. 1-21.

[59]. P. Sharma. L. Lao. H. Yeung, Clamp-on microwave – A new tool for flow diagnostics, in *Proceedings of the 2nd Annual Conference of Flow Measurement Institute,* Coventry, UK, 2016.

[60]. H. Seraj, M. Fua, Review of water salinity measurement methods and considering salinity in measuring water area phase fraction of wet gas, *Sensors and Transducers*, Vol. 162, Issue 1, 2014, pp. 208-214.

[61]. B. N. Scott, High Water Cut Well Measurements Using Heuristic Salinity Determination, USPTO, Patent US7587290B2, *USA*, 2009.

[62]. M. J. Black, M. Alhusseini, M. N. Noui-mehidi, High salinity permittivity models for water cut sensing, *IEEE Transactions on Instrumentation and Measurement*, Vol. 62, Issue 10, 2013, pp. 2805-2811.

[63]. C. Saetre, G. A. Johansen, S. A. Tjugum, Salinity and flow regime independent multiphase flow measurements, *Flow Measurement and Instrumentation*, Vol. 21, Issue 4, 2010, pp. 454-461.

[64]. R. Somaraju, J. Trumpf, Frequency, temperature and salinity variation of the permittivity of seawater, *IEEE Transactions on Antennas and Propagation*, Vol. 54, Issue 11, 2006, pp. 3441-3448.

[65]. P. Sharma, H. Yeung, A non-intrusive sensor for water-in-oil monitoring in multiphase flow, in *Proceedings of the 9th International Symposium on Measurement Techniques for Multiphase Flow*, Sapporo, Hokkaido, Japan, 2015.

[66]. F. Viana, *et al.*, Challenges of multiphase flow metering in heavy oil applications, in *Proceedings of SPE Heavy Oil Conference*, Canada, 2013, pp. 11-13.

Chapter 5

Modelling Methods for the Colored Noise of Inertial Sensors

Kedong Wang

5.1. Introduction

The inertial sensors, both gyroscope and accelerometer, have errors in their output inevitable [1]. The errors include the deterministic and the random. In the theory, the former error can be compensated for completely, but the latter cannot be at all. Therefore, the performance of the sensors will be determined by the residual random error significantly if the deterministic error has been compensated for by calibration to the most extent.

Although it is convenient for applications to assume the random error as a white noise, it is usually a colored one actually. Hence, many efforts have been devoted to the random error modelling as a colored noise [1-3]. There are three popular methods to model the colored noise up to now at least. The first is the shaping filter based on the spectral analysis [4]. As long as the power spectral density (PSD) of the colored noise is rational and even, a linear shaping filter with an input of a white noise can be derived for the colored noise. Of course, if the PSD of the colored noise is not rational, it is impossible to derive the shaping filter any more. The second is the auto-regressive moving average (ARMA) method based on time series [2]. If the sequence of the colored noise is stationary, it can be modelled as an ARMA one. If the sequence is not stationary, it can be differenced by several times to obtain a stationary one. That is, the ARMA model is very adaptive in implementation, but there is no exact meaning of the model. Therefore, engineers and manufacturers would like to use the third method, the Allan variance, to model the

Kedong Wang
School of Astronautics, Beihang University, Beijing, China

colored noise [1]. The colored noise is modelled as the superimposition of several typical noises, such as quantization noise, angular random walk, bias instability, drift rate ramp, and etc., so that the modelled noise is meaningful in concept. In this chapter, the three methods will be introduced with examples.

5.2. White Noise and Colored Noise

A white noise is defined as

$$\Phi_{xx}(\omega) = \Phi_0, \tag{5.1}$$

where $\Phi_{xx}(\omega)$ is the PSD of the noise x, $-\infty < \omega < \infty$, and Φ_0 is a constant greater than 0. According to (5.1), the white noise is an ideal model since its power will be infinite but the real noise always has a finite power.

The autocorrelation function (ACF) of the white noise can be obtained by Fourier transform (FT). That is

$$R_{xx}(\tau) = \frac{1}{2\pi} \int_{-\infty}^{\infty} \Phi_0 \, e^{j\omega\tau} \, d\omega = \Phi_0 \delta(\tau), \tag{5.2}$$

where $R_{xx}(\tau)$ is the ACF of x, τ is the correlation time, and $\delta(\tau)$ is the Dirac δ function. According to (5.2), the following points can be derived. i) The white noise can also not be implemented in time domain since the Dirac δ function is ideal. ii) The ACF of the white noise is infinite when $\tau = 0$ while it is zero when $\tau \neq 0$, which means that the white noise is totally uncorrelated when $\tau \neq 0$. Hence, the white noise cannot be predicted by its past state.

The noises which do not abide by the definition in (5.1) are called as the colored ones. As analyzed above, all the real noises are colored. However, most of the colored noises can be modelled as the outputs of the linear systems with the input of a white noise. Derivation of the linear systems for the colored noises is called as the whiten process or the colored noise modelling.

The popular methods to model a colored noise include the shaping filter based on PSD, ARMA based on time series, and Allan variance, which will be introduced hereinafter in detail.

5.3. Shaping Filter

In a linear system, its output PSD can be derived as

$$\Phi_{yy}(\omega) = |h(j\omega)|^2 \Phi_{xx}(\omega), \tag{5.3}$$

where $\Phi_{yy}(\omega)$ is the PSD of the output y, $h(j\omega)$ is the unit impulse response function of the linear system, and $j^2 = -1$. If the input $x(t)$ is a white noise with the PSD of 1, $\Phi_{yy}(\omega) = |h(j\omega)|^2 = h(j\omega)h^*(j\omega)$ where $h^*(j\omega)$ is the conjugate complex of $h(j\omega)$. Hence, if the PSD of a colored noise, $\Phi_{yy}(\omega)$, can be rationally decomposed as $\phi_{yy}(j\omega)\phi_{yy}(-j\omega)$, $h(j\omega) = \phi_{yy}(j\omega)$ which is the transfer function of the shaping filter. If $y(t)$ is ergodic, $\Phi_{yy}(\omega)$ will be even and can be rationally decomposed as $\phi_{yy}(j\omega)\phi_{yy}(-j\omega)$ where $\phi_{yy}(s)$ is the transfer function $h(s)$ and s is the Laplace operator, which can be proven as follows.

For an ergodic process $y(t)$, $R_{yy}(\tau) = R_{yy}(-\tau)$ and there is

$$
\begin{aligned}
\Phi_{yy}(\omega) &= \int_{-\infty}^{\infty} R_{yy}(\tau)e^{-j\omega\tau}d\tau = \\
&= \int_{-\infty}^{\infty} R_{yy}(\tau)\cos\omega\tau\,d\tau - j\int_{-\infty}^{\infty} R_{yy}(\tau)\sin\omega\tau\,d\tau = \tag{5.4} \\
&= 2\int_{0}^{\infty} R_{yy}(\tau)\cos\omega\tau\,d\tau
\end{aligned}
$$

In (5.4), $\Phi_{yy}(\omega)$ is even. If $\Phi_{yy}(\omega)$ is rational, it can be assumed as

$$
\begin{aligned}
\Phi_{yy}(\omega) &= |\phi_{yy}(\omega)|^2 = \frac{b_0\omega^{2m} + b_1\omega^{2m-2} + \cdots + b_m}{a_0\omega^{2n} + a_1\omega^{2n-2} + \cdots + a_n} = \\
&= \frac{b_0(\omega^2 - z_1)(\omega^2 - z_2)\cdots(\omega^2 - z_m)}{a_0(\omega^2 - p_1)(\omega^2 - p_2)\cdots(\omega^2 - p_n)} = \\
&= \sqrt{\frac{b_0}{a_0}}\frac{(\sqrt{-z_1}+j\omega)(\sqrt{-z_2}+j\omega)\cdots(\sqrt{-z_m}+j\omega)}{(\sqrt{-p_1}+j\omega)(\sqrt{-p_2}+j\omega)\cdots(\sqrt{-p_n}+j\omega)}.
\end{aligned}
$$

$$\bullet \sqrt{\frac{b_0}{a_0}} \frac{\left(\sqrt{-z_1}-j\omega\right)\left(\sqrt{-z_2}-j\omega\right)\cdots\left(\sqrt{-z_m}-j\omega\right)}{\left(\sqrt{-p_1}-j\omega\right)\left(\sqrt{-p_2}-j\omega\right)\cdots\left(\sqrt{-p_n}-j\omega\right)} \tag{5.5}$$
$$= \phi_{yy}\left(j\omega\right)\phi_{yy}\left(-j\omega\right),$$

where $a_0 \neq 0$, $b_0 \neq 0$, $m < n$ and

$$\phi_{yy}\left(j\omega\right) = \sqrt{\frac{b_0}{a_0}} \frac{\left(\sqrt{-z_1}+j\omega\right)\left(\sqrt{-z_2}+j\omega\right)\cdots\left(\sqrt{-z_m}+j\omega\right)}{\left(\sqrt{-p_1}+j\omega\right)\left(\sqrt{-p_2}+j\omega\right)\cdots\left(\sqrt{-p_n}+j\omega\right)} \tag{5.6}$$

Example 1

The ACF of a stationary and ergodic random process $x(t)$ is $R_{xx}\left(\tau\right) = k\,e^{-a|\tau|}\,(a > 0)$. Try to derive its discrete model.

The PSD of $x(t)$ can be derived as

$$\Phi_{xx}\left(\omega\right) = \int_{-\infty}^{\infty} R_{xx}\left(\tau\right)e^{-j\omega\tau}\,d\tau = \int_{-\infty}^{\infty} k\,e^{-a|\tau|}\,e^{-j\omega\tau}\,d\tau =$$
$$= k\left(\int_{-\infty}^{0} k\,e^{a\tau}\,e^{-j\omega\tau}\,d\tau + \int_{0}^{\infty} k\,e^{-a\tau}\,e^{-j\omega\tau}\,d\tau\right) =$$
$$= k\left(\frac{1}{a - j\omega} + \frac{1}{a + j\omega}\right) = \frac{2ka}{\omega^2 + a^2} = \frac{\sqrt{2ka}}{a + j\omega}\frac{\sqrt{2ka}}{a - j\omega}$$

Hence, the transfer function of the shaping filter is

$$h(s) = \frac{\sqrt{2ka}}{a + s} = \frac{X(s)}{W(s)}$$

The corresponding differential function of the shaping function is

$$\dot{x}(t) = -ax(t) + \sqrt{2ka}w(t),$$

where $w(t)$ is a white noise with zero mean and the PSD of 1. Obviously, the first and the second moments of $w(t)$ are

$$\begin{cases} E\left[\sqrt{2ka}w(t)\right] = 0 \\ E\left[\sqrt{2ka}w(t)\sqrt{2ka}w(\tau)\right] = 2ka\delta(t - \tau) \end{cases},$$

where $E(\cdot)$ is the expectation. The discrete form of the shaping filter is

$$x_{k+1} = e^{-a(t_{k+1}-t_k)}x_k + \int_{t_k}^{t_{k+1}} e^{-a(t_{k+1}-\tau)}\sqrt{2ka}w(\tau)d\tau = \Phi x_k + w_k,$$

where

$$\begin{cases} E(w_k) = \int_{t_k}^{t_{k+1}} e^{-a(t_{k+1}-\tau)}\sqrt{2ka}E[w(\tau)]d\tau = 0 \\ E(w_k w_j) = E\left[\int_{t_k}^{t_{k+1}} e^{-a(t_{k+1}-\tau)}\sqrt{2ka}w(\tau)d\tau \int_{t_j}^{t_{j+1}} e^{-a(t_{j+1}-\lambda)}\sqrt{2ka}w(\lambda)d\lambda\right] \\ = 2ka\int_{t_k}^{t_{k+1}}\int_{t_j}^{t_{j+1}} e^{-a(t_{k+1}-\tau)}e^{-a(t_{j+1}-\lambda)}E[w(\tau)w(\lambda)]d\lambda d\tau \\ = 2ka\int_{t_k}^{t_{k+1}}\int_{t_j}^{t_{j+1}} e^{-a(t_{k+1}-\tau)}e^{-a(t_{j+1}-\lambda)}\delta(\tau-\lambda)d\lambda d\tau \\ = 2ka\int_{t_k}^{t_{k+1}} e^{-2a(t_{k+1}-\tau)}\delta_{kj}d\tau = k\left[1-e^{-2a(t_{k+1}-t_k)}\right]\delta_{kj} \end{cases}$$

According to the above derivation, the model of $x(t)$ is a first-order Markov process. Fig. 5.1 illustrates the block diagram of the model.

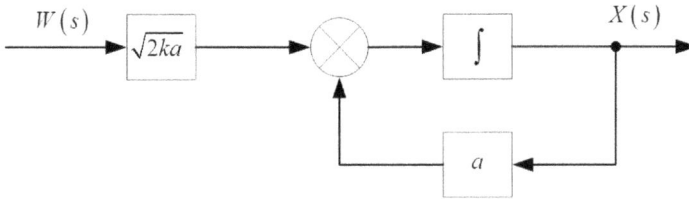

Fig. 5.1. Block diagram of the shaping function.

5.4. ARMA

5.4.1. Definition

A stationary discrete colored noise $\{x_k\}$ can be modeled as

$$x_k = \sum_{i=1}^{p}\phi_i x_{k-i} + w_k - \sum_{i=1}^{q}\theta_i w_{k-i}, \tag{5.7}$$

where $\phi_i\,(i=1,2,\cdots,p)$ is the auto-regressive (AR) parameter, $\theta_i\,(i=1,2,\cdots,q)$ is the moving average (MA) parameter, and $\{w_k\}$ is a zero mean white noise sequence. The model is called as an $\mathrm{ARMA}(p,q)$ one. If $p=0$, it is called as a q order $\mathrm{MA}(q)$ one. If $q=0$, it is called as a p order $\mathrm{AR}(p)$ one.

If the sequence is not stationary, i.e., its statistics are time related, it has to be differenced m times to obtain a stationary one before it is modeled as an ARMA one. The ARMA model by differencing is called as an auto-regressive integrated moving average (ARIMA) one. An $\mathrm{ARIMA}(p,m,q)$ model is

$$
\begin{cases}
y_k = x_k - x_{k-m} \\
y_k = \displaystyle\sum_{i=1}^{p} \phi_i y_{k-i} + w_k - \sum_{i=1}^{q} \theta_i w_{k-i}
\end{cases}
\tag{5.8}
$$

where $\{y_k\}$ is the non-stationary sequence.

If there are the r order interferences in an $\mathrm{ARMA}(p,q)$ sequence, the model is called as an auto-regressive moving average model with exogenous inputs, or $\mathrm{ARMAX}(p,r,q)$.

$$
\begin{cases}
y_k = \displaystyle\sum_{i=1}^{r} a_i y_{k-i} + u_k - \sum_{i=1}^{r} b_i u_{k-i} + e_k \\
e_k = \displaystyle\sum_{i=1}^{p} \phi_i e_{k-i} + w_k - \sum_{i=1}^{q} \theta_i w_{k-i}
\end{cases}
\tag{5.9}
$$

where $\{u_k\}$ is the interference sequence and b_i is its coefficient.

Herein, only the stationary colored noise sequence is taken into account. Its ARMA modeling is to estimate the ARMA parameters including p, q, ϕ_i, θ_i and the variance of w_k, which will be detailed in the following.

5.4.2. Stationary, Invertible, and Expandable

5.4.2.1. Stationary

(5.7) is rewritten as

$$\begin{cases} \phi\left(d^{-1}\right)x_k = \theta\left(d^{-1}\right)w_k \\ \phi\left(d^{-1}\right) = 1 - \phi_1 d^{-1} - \cdots - \phi_p d^{-p}, \\ \theta\left(d^{-1}\right) = 1 - \theta_1 d^{-1} - \cdots - \theta_q d^{-q} \end{cases} \qquad (5.10)$$

where d^{-1} is the delay operator. It can be proven that the sufficient condition of a stationary ARMA process is the zeros of $\phi(x)$ locating at the outside of a unit circle, but the proof is omitted.

Example 2

An AR(1) model is

$$x_k = \phi x_{k-1} + w_k,$$

where w_k is the zero mean white noise with the variance of σ_w^2. The sufficient condition to ensure it stationary can be derived as follows.

The above model can be rewritten as

$$x_k = \frac{1}{1 - \phi d^{-1}} w_k$$

The series expansion of the model is

$$x_k = \sum_{i=0}^{\infty} \left(\phi d^{-1}\right)^i w_k = \sum_{i=0}^{\infty} \left(\phi^i w_{k-i}\right)$$

Obviously, only if

$$\sum_{i=0}^{\infty} \phi^{2i} < \infty$$

The sequence $\{x_k\}$ is stationary and convergent in the merit of mean square error (MSE), so the inequality is the sufficient condition to ensure the AR(1) sequence stationary. The equivalent form of the inequality is $|\phi| < 1$, i.e., the zeros of the polynomial $\phi\left(d^{-1}\right) = 1 - \phi d^{-1}$ locate at the outside of a unit circle.

5.4.2.2. Invertible

Similarly, the sufficient condition to ensure an ARMA model invertible is the zeros of the polynomial of $\theta(x)$ locating outside a unit circle.

Example 3

An MA model is

$$x_k = w_k - \theta w_{k-1},$$

where w_k is the zero mean white noise with the variance of σ_w^2. The sufficient condition to ensure it invertible can be derived as follows.

The model can be rewritten as

$$w_k = \frac{1}{1 - \theta d^{-1}} x_k = \sum_{i=0}^{\infty} \left(\theta^i x_{k-i} \right)$$

The sufficient and necessary condition to ensure the series convergent in the merit of MSE is that when $m \to \infty$ and $n \to \infty$, there is

$$\mathrm{E}\left[\sum_{i=m}^{n} \left(\theta^i x_{k-i} \right) \right]^2 = \mathrm{E}\left[\sum_{i=m}^{n} \left(\theta^i x_{k-i} \right) \sum_{j=m}^{n} \left(\theta^i x_{k-j} \right) \right] =$$

$$= \sum_{i=m}^{n} \sum_{j=m}^{n} \left[\theta^i \theta^j R_{xx} (i-j) \right] \to 0$$

Since

$$R_{xx}(\tau) = \mathrm{E}(x_k x_{k-\tau}) = \mathrm{E}\left[(w_k - \theta w_{k-1})(w_{k-\tau} - \theta w_{k-\tau-1}) \right]$$

$$= \begin{cases} (1+\theta^2)\sigma_w^2, & \tau = 0 \\ -\theta\sigma_w^2, & \tau = 1 \\ 0, & \tau > 1 \end{cases}$$

Hence, if both θ and σ_w^2 are bounded, $R_{xx}(\tau)$ is also bounded so that the convergent condition of the series is $|\theta| < 1$, i.e., the zeros of the polynomial $\theta(x)$ locating at the outside of a unit circle.

5.4.2.3. Expandable

In the above two examples, the series expansion of an ARMA model has been used. Herein, the way to expand the ARMA model will be provided if it is expandable.

If the ARMA model in (5.10) is stationary, it can be expanded as

$$x_k = \frac{\theta(d^{-1})}{\phi(d^{-1})} w_k = \sum_{i=0}^{\infty} \varphi_i d^{-i} w_k = \sum_{i=0}^{\infty} \varphi_i w_{k-i}, \tag{5.11}$$

where

$$\frac{\theta(d^{-1})}{\phi(d^{-1})} = \sum_{i=0}^{\infty} \varphi_i d^{-i} \tag{5.12}$$

That is

$$1 - \theta_1 d^{-1} - \cdots - \theta_q d^{-q} = \left(1 - \phi_1 d^{-1} - \cdots - \phi_p d^{-p}\right) \sum_{i=0}^{\infty} \varphi_i d^{-i} \tag{5.13}$$

Hence

$$\begin{cases} \varphi_0 = 1 \\ \varphi_1 = \phi_1 - \theta_1 \\ \vdots \\ \varphi_i = \phi_1 \varphi_{i-1} + \phi_2 \varphi_{i-2} + \cdots + \phi_p \varphi_{i-p} - \theta_i \end{cases} . \tag{5.14}$$

Obviously, when $i > q$, $\theta_i = 0$ and there is

$$\varphi_i - \left(\phi_1 \varphi_{i-1} + \phi_2 \varphi_{i-2} + \cdots + \phi_p \varphi_{i-p}\right) = \phi(d^{-1}) \varphi_i = 0 \tag{5.15}$$

Similarly, if the ARMA model is invertible, there is

$$\begin{cases} w_k = \frac{\phi(d^{-1})}{\theta(d^{-1})} x_k = x_k - \sum_{i=1}^{\infty} \vartheta_i d^{-i} x_k = x_k - \sum_{i=1}^{\infty} \vartheta_i x_{k-i} \\ \vartheta_i = \theta_1 \vartheta_{i-1} + \theta_2 \vartheta_{i-2} + \cdots + \theta_q \vartheta_{i-q} + \phi_i \end{cases} , \tag{5.16}$$

where $\vartheta_i = 0 (i < 0)$ and $\phi_i = 0 (i > p)$.

According to (5.11) and (5.16), if the ARMA model is stationary and invertible, it can be equivalent to an MA(∞) or an AR(∞) one. Of course, the limited items will be kept in the series to achieve the expected accuracy since it is impossible to expand the model with the infinite items in implementation.

Example 4

An ARMA model is

$$x_k = \phi x_{k-1} + w_k - \theta w_{k-1},$$

where $\phi = 0.2$, $\theta = 0.5$, $E(w_k) = 0$, $E(w_k w_j) = \sigma_w^2 \delta_{kj}$, δ_{kj} is the Kronecker δ function ($\delta_{kj} = 1$ if $k = j$, otherwise $\delta_{kj} = 0$), and $\sigma_w^2 = 1$. It can be expanded as an AR or an MA model as follows.

Since the ARMA(1,1) model is stationary and invertible, it can be expanded as an AR or an MA model.

Firstly, it can be expanded as an AR model as follows. Since

$$\begin{cases} \vartheta_0 = -1 \\ \vartheta_1 = \phi - \theta \\ \vartheta_2 = \theta \vartheta_1 = \theta(\phi - \theta) \\ \vdots \\ \vartheta_i = \theta^{i-1}(\phi - \theta) \end{cases}$$

There is

$$x_k = \sum_{i=1}^{\infty} \theta^{i-1}(\phi - \theta) x_{k-i} + w_k$$

Then, it can be expanded as an MA model similarly as

$$x_k = \sum_{i=1}^{\infty} \phi^{i-1}(\phi - \theta) w_{k-i} + w_k$$

Fig. 5.2 depicts the difference between the expanded MA with the first 30 items and the original ARMA sequence. The accuracy of the AR expansion is similar. The effective method to improve the expansion accuracy is keeping more items.

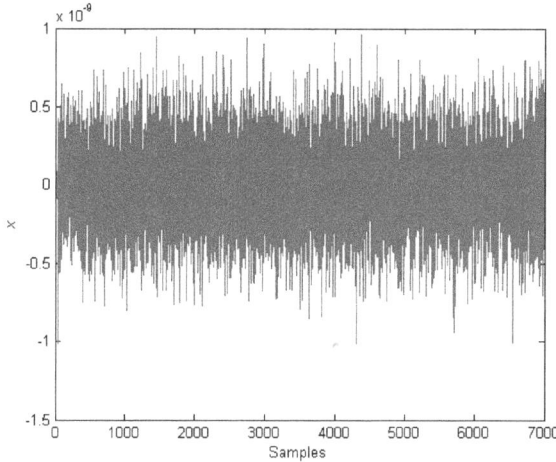

Fig. 5.2. The difference between the original ARMA(1,1) sequence
and the MA expansion with the first 30 items.

5.4.3. ARMA Modeling

In the ARMA modeling, the AR order, the MA order, the AR parameters,
the MA parameter, and σ_w^2 have to be estimated. Usually, both the AR
and the MA orders should be estimated firstly, but they are introduced
finally herein. Hence, the AR and the MA parameters and the variance
of noise σ_w^2 will be estimated with the assumption of the known AR and
MA orders in the following. It should be noted that there are many
estimation methods to determine the ARMA model parameters and the
following estimation method is just one of them. The readers can refer
to the related literature [5-6].

5.4.3.1. Estimation of the AR Parameters

Multiply $x_{k-\tau}$ at both sides of (5.10) and derive the expectation as

$$E\left(x_k x_{k-\tau}\right) = R_{xx}\left(\tau\right) = \sum_{i=1}^{p} \phi_i R_{xx}\left(\tau - i\right) + E\left(w_k x_{k-\tau}\right) -$$
$$- \sum_{j=1}^{q} \theta_j E\left(w_{k-j} x_{k-\tau}\right) \tag{5.17}$$

If $\tau > q$, there is

$$R_{xx}(\tau) = \sum_{i=1}^{p} \phi_i R_{xx}(\tau - i) \qquad (5.18)$$

That is, the ACF of the sequence just relates to the AR parameters, so the AR parameters can be estimated as

$$
\begin{bmatrix}
R_{xx}(q) & R_{xx}(q-1) & \cdots & R_{xx}(q-p+1) \\
R_{xx}(q+1) & R_{xx}(q) & \cdots & R_{xx}(q-p+2) \\
\vdots & \vdots & \cdots & \vdots \\
R_{xx}(s-1) & R_{xx}(s-2) & \cdots & R_{xx}(s-p)
\end{bmatrix}
\begin{bmatrix}
\phi_1 \\
\phi_2 \\
\vdots \\
\phi_p
\end{bmatrix}
$$

$$
=
\begin{bmatrix}
R_{xx}(q+1) \\
R_{xx}(q+2) \\
\vdots \\
R_{xx}(s)
\end{bmatrix}
\qquad (5.19)
$$

(5.19) is called as the Yule-Walk equation. Let

$$
\mathbf{R}_{xx} =
\begin{bmatrix}
R_{xx}(q) & R_{xx}(q-1) & \cdots & R_{xx}(q-p+1) \\
R_{xx}(q+1) & R_{xx}(q) & \cdots & R_{xx}(q-p+2) \\
\vdots & \vdots & \cdots & \vdots \\
R_{xx}(s-1) & R_{xx}(s-2) & \cdots & R_{xx}(s-p)
\end{bmatrix}
$$

$$
\mathbf{R}_s = \begin{bmatrix} R_{xx}(q+1) & R_{xx}(q+2) & \cdots & R_{xx}(s) \end{bmatrix}^{\mathrm{T}}
$$

$$
\boldsymbol{\varphi} = \begin{bmatrix} \phi_1 & \phi_2 & \cdots & \phi_p \end{bmatrix}^{\mathrm{T}}
$$

Rewrite (5.19) as

$$\mathbf{R}_s = \mathbf{R}_{xx}\boldsymbol{\varphi} \qquad (5.20)$$

The least square estimation of $\boldsymbol{\varphi}$ is

$$\hat{\boldsymbol{\varphi}} = \left(\mathbf{R}_{xx}^{\mathrm{T}} \mathbf{R}_{xx} \right)^{-1} \mathbf{R}_{xx}^{\mathrm{T}} \mathbf{R}_s \qquad (5.21)$$

5.4.3.2. Estimation of the MA Parameters and the Variance of Noise

After estimating the AR parameters, the original sequence can be filtered by the estimated AR model. The filtered sequence will approach to an

MA process, so the estimation of the MA parameters is realized by the filtered residual.

Assume the filtered residual as

$$r_k = w_k - \theta_1 w_{k-1} - \cdots - \theta_q w_{k-q}, \tag{5.22}$$

where $E(w_k) = 0$ and $E(w_k w_j) = \sigma_w^2 \delta_{kj}$. The following task is to estimate $\theta_i (i = 1, 2, \cdots, q)$ and σ_w^2.

Assume

$$R_{rr}(\tau) = \begin{cases} E(r_k r_{k-\tau}), & \tau = 0, 1, \cdots, q \\ 0, & \tau > q \end{cases} \tag{5.23}$$

Let

$$\begin{cases} R_{rw}(\tau) = E(r_k w_{k-\tau}) \\ R_{ww}(0) = E(w_k w_k) = \sigma_w^2 \end{cases} \tag{5.24}$$

According to (5.22), there is

$$E(r_k r_{k-\tau}) = -\theta_\tau \sigma_w^2 \tag{5.25}$$

That is

$$\theta_\tau = -\frac{E(r_k r_{k-\tau})}{\sigma_w^2} \tag{5.26}$$

Hence, $\theta_i (i = 1, 2, \cdots, q)$ and σ_w^2 can be derived by (5.24) and (5.26), but they are usually estimated recursively to achieve the more accurate estimation. Their estimation is as follows.

According to (5.22), there is

$$\begin{cases} r_{k-m} = w_{k-m} - \theta_1 w_{k-m-1} - \cdots - \theta_q w_{k-m-q} \\ w_{k-m} = r_{k-m} + \theta_1 w_{k-m-1} + \cdots + \theta_q w_{k-m-q} \end{cases} \tag{5.27}$$

We can get

$$E\left(r_{k-m}w_{k-m-\tau}\right) = -\theta_\tau E\left(w_{k-m-\tau}w_{k-m-\tau}\right) \tag{5.28}$$

or

$$\theta_\tau = -\frac{E\left(r_{k-m}w_{k-m-\tau}\right)}{E\left(w_{k-m-\tau}w_{k-m-\tau}\right)} \tag{5.29}$$

Furthermore

$$R_{rw}\left(k,k-m\right) = E\left(r_k w_{k-m}\right)$$

$$= E\left\{r_k\left[r_{k-m} - \sum_{i=1}^{q}\frac{E\left(r_{k-m}w_{k-m-i}\right)}{E\left(w_{k-m-i}w_{k-m-i}\right)}w_{k-i-m}\right]\right\}$$

$$= R_{rr}\left(m\right) - \sum_{i=1}^{q-m}\frac{R_{rw}\left(k-m,k-m-i\right)R_{rw}\left(k,k-i-m\right)}{R_{ww}\left(k-m-i,k-m-i\right)}$$

$$= R_{rr}\left(m\right) - \sum_{j=m+1}^{j=m+i}\frac{R_{rw}\left(k,k-j\right)R_{rw}\left(k-m,k-j\right)}{R_{ww}\left(k-j,k-j\right)} \tag{5.30}$$

In the derivation of (5.30), it is used that if $j > q$, $R_{rw}\left(k,k-j\right) = 0$. Iterate (5.29) for many times to the converged results which can be viewed as the final estimation of the MA parameters.

The above estimation can be summarized as

$$\begin{cases} \theta_i = -\lim_{k\to\infty}\dfrac{R_{rw}\left(k,k-i\right)}{R_{rw}\left(k,k\right)}, i = 1,2,\cdots,q \\[2mm] \sigma_w^2 = \lim_{k\to\infty}R_{rw}\left(k,k\right) \\[2mm] R_{rw}\left(k,k-i\right) = R_{rr}\left(i\right) - \displaystyle\sum_{j=i+1}^{q}\dfrac{R_{rw}\left(k,k-j\right)R_{rw}\left(k-i,k-j\right)}{R_{ww}\left(k-j,k-j\right)} \end{cases}, \tag{5.31}$$

where $k = 0,1,2,\cdots,N$; $m = k,k-1,\cdots,0$ and

$$\begin{cases} R_{rw}\left(0,0\right) = R_{rw}\left(0\right) \\[2mm] R_{rw}\left(k,k-s\right) = 0, \quad k < s \\[2mm] \dfrac{1}{R_{rw}\left(k-s,k-s\right)} = 0, \quad k < s \end{cases} \tag{5.32}$$

The estimation by (5.31) and (5.32) is usually called as the Gevers-Wouters algorithm or the G-W one for brief [7].

Example 5

A sequence abided by the following ARMA model as

$$x_k + 0.579x_{k-1} + 0.442x_{k-2} - 0.769x_{k-3} = w_k + 0.494w_{k-1} - 0.297w_{k-2},$$

where $E(w_k) = 0$, $E(w_k w_j) = \sigma_w^2 \delta_{kj}$, and $\sigma_w^2 = 1$. Fig. 5.3 depicts one implementation of the sequence. The model parameters can be estimated by the implementation as follows.

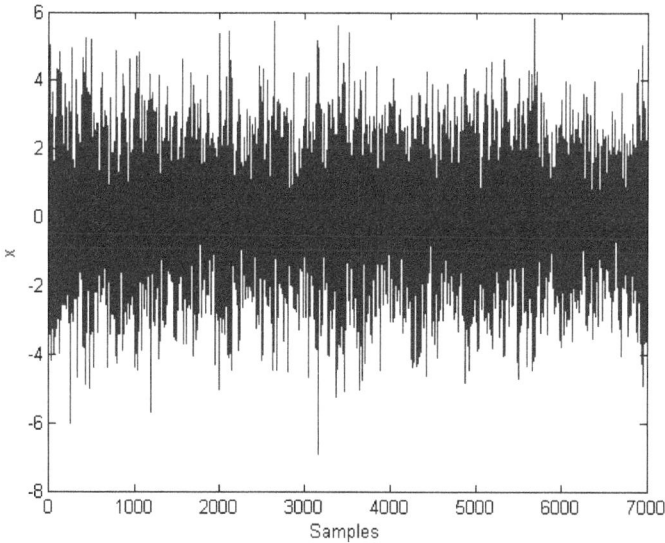

Fig. 5.3. An implementation of the sequence.

It is known that $p = 3$ and $q = 2$. We need to estimate the AR parameters, the MA parameters, and σ_w^2.

The AR parameters are estimated firstly herein. Let $s = 3p$ in (5.20) where

$$\mathbf{R}_s = \begin{bmatrix} -1.2438 & -0.2182 & -1.807 & -0.1785 & -0.7438 & 0.8944 & 0.3254 \end{bmatrix}^{\mathrm{T}}$$

191

$$\mathbf{R}_{xx} = \begin{bmatrix} 1.4686 & 0.1433 & 2.8214 \\ -1.2438 & 1.4686 & 0.1433 \\ -0.2182 & -1.2438 & 1.4686 \\ -1.807 & -0.2182 & -1.2438 \\ -0.1785 & -1.807 & -0.2182 \\ -0.7438 & -0.1785 & -1.807 \\ 0.8944 & -0.7438 & -0.1785 \end{bmatrix}$$

Substitute the above values into (5.21) and obtain $\hat{\boldsymbol{\varphi}} = \begin{bmatrix} 0.5868 & 0.4383 & -0.7732 \end{bmatrix}^{\mathrm{T}}$.

Then, the MA parameters and σ_w^2 will be estimated. The estimated AR parameters will be used to construct a filter to process the original sequence. The residual of the filtered sequence is shown in Fig. 5.4. The estimation of the MA parameters and σ_w^2 is based on the residual according to (5.31) and (5.32). The iterative results after 100 iterations are $\hat{\theta}_1 = -0.4955$, $\hat{\theta}_2 = 0.268$, and $\hat{\sigma}_w^2 = 0.9765$.

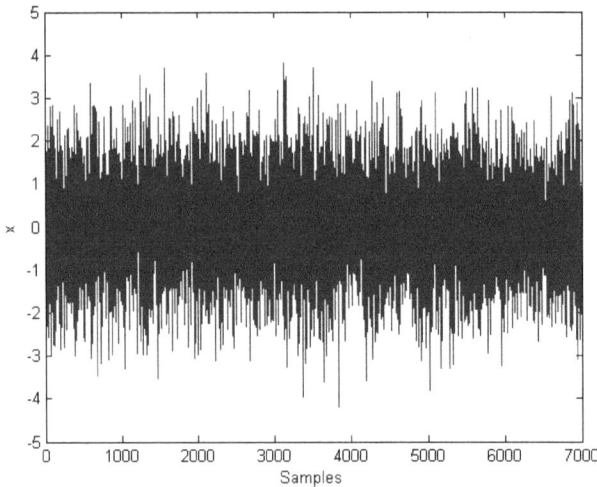

Fig. 5.4. The residual of the filtered sequence.

Compared with the model's real parameters, the estimated ones are very close to their real values, which proves the effectiveness of the method.

5.4.3.3. Estimation of the AR and the MA Orders

In the above estimation, the AR and the MA orders are assumed as known, but they are unknown in practice while what we have is just the samples of a sequence. Hence, it is necessary to estimate the model orders before the other parameters' estimation. The popular methods to estimate the model orders include final prediction error (FPE), Akaike information criterion (AIC), minimum description length (MDL), and etc. [8] Herein, we will introduce MDL.

Construct the following equation according to the samples of a sequence as

$$\mathbf{D}\psi = \upsilon, \tag{5.33}$$

where

$$\mathbf{D} = \begin{bmatrix} x(1) & 0 & \cdots & 0 & w(1) & 0 & \cdots & 0 \\ x(2) & x(1) & \cdots & 0 & w(2) & w(1) & \cdots & 0 \\ \vdots & \vdots & \cdots & \vdots & \vdots & \vdots & \cdots & \vdots \\ x(N) & x(N-1) & \cdots & x(N-p) & w(N) & w(N-1) & \cdots & w(N-q) \end{bmatrix}$$

$$\psi = \begin{bmatrix} \phi_0 & -\phi_1 & \cdots & -\phi_p & -\theta_0 & \theta_1 & \cdots & \theta_q \end{bmatrix}^{\mathrm{T}}$$

$$\upsilon = \begin{bmatrix} v(1) & v(2) & \cdots & v(N) \end{bmatrix}^{\mathrm{T}}$$

In fact, (5.33) is derived by the left side of (5.10) minus the right side. The residual υ is assumed as a zero mean white noise with the variance of σ_v^2. Assume $\phi_0 = \theta_0 = 1$. The following matrix can be derived by (5.33)

$$\mathbf{M} = \mathbf{D}^{\mathrm{T}}\mathbf{D} \tag{5.34}$$

In the derivation of $\mathbf{M}$, $w(i)(i = 1, 2, \cdots, N)$ will be used, but it is unknown in the estimation. Hence, its value has to be estimated herein firstly. It is assumed that the sequence is modelled as a high-order MA process. According (5.16), there is

$$w(i) \approx x(i) - \sum_{j=1}^{H_q} \vartheta_j x_{i-j} = x(i) - \mathbf{x}^{\mathrm{T}}(i)\boldsymbol{\alpha}, \tag{5.35}$$

where H_q is the MA order, $\mathbf{x}(i) = \begin{bmatrix} x_{i-1} & x_{i-2} & \cdots & x_{i-H_q} \end{bmatrix}^{\mathrm{T}}$, and $\boldsymbol{\alpha} = \begin{bmatrix} \vartheta_1 & \vartheta_2 & \cdots & \vartheta_{H_q} \end{bmatrix}^{\mathrm{T}}$. The least square estimation of $\boldsymbol{\alpha}$ is

$$\hat{\boldsymbol{\alpha}} = \left[\frac{1}{N+1} \sum_{i=1}^{N} \mathbf{x}(i) \mathbf{x}^{\mathrm{T}}(i) \right]^{-1} \frac{1}{N+1} \sum_{i=1}^{N} \mathbf{x}(i) x(i) \qquad (5.36)$$

After deriving $\hat{\boldsymbol{\alpha}}$, the estimated $w(i)$, $\hat{w}(i)$, can be obtained in (5.35). $\hat{w}(i)$ will be used in (5.34) to calculate $\mathbf{M}$.

In (5.34), the eigenvalue of $\mathbf{M}$ can be calculated when p and q are known. Assume $\lambda_{\min}(p,q)$ is the minimum eigenvalue. Obviously, there is a value of $\lambda_{\min}(p,q)$ for a pair of p and q. In a range of p and q, such as $[0,10]$, a matrix of $\lambda_{\min}(p,q)$, $\mathbf{J}(p,q)$, will be acquired. In the matrix, the minimum values of $\dfrac{\lambda_{\min}(p,q)}{\lambda_{\min}(p-1,q)}$ and $\dfrac{\lambda_{\min}(p,q)}{\lambda_{\min}(p,q-1)}$ will be found and the corresponding p and q will be marked since they are the estimation. The MDL method is derived by the maximum likelihood estimation, which is referred to the literature [8].

Example 6

The samples of the sequence are the same as Example 5. The model orders can be estimated as follows if an ARMA model is assumed.

The number of the samples is $N = 7000$. Assume $H_q = 100$. Calculate $\hat{\boldsymbol{\alpha}}$ according to (5.36). The values of $\hat{w}(i)$ can be obtained according to (5.35) further, which is depicted in Fig. 5.5. Then, change the values of p and q in the range of $[0, 10]$ to acquire the value of $\lambda_{\min}(p,q)$ of the matrix $\mathbf{M}$ in (5.34). After traversing all the values of p and q, the matrix $\mathbf{J}(p,q)$ listed in Table 5.1 can be obtained. The values of $\dfrac{\lambda_{\min}(p,q)}{\lambda_{\min}(p-1,q)}$ and $\dfrac{\lambda_{\min}(p,q)}{\lambda_{\min}(p,q-1)}$ listed in Tables 5.2 and 5.3 can be calculated further according to $\mathbf{J}(p,q)$.

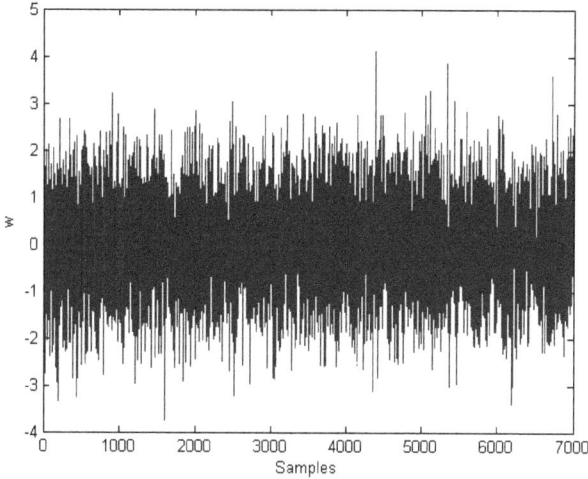

Fig. 5.5. The estimated values of $w(i)$.

Table 5.1. The values of the minimum eigenvalues.

q\p	0	1	2	3	4	5	6	7	8	9	10
0	3859.8087	520.3789	501.8985	200.0955	27.2485	21.9775	19.4714	17.7881	16.3948	15.3951	14.7832
1	3845.7109	775.3876	589.3162	86.9750	26.9234	24.3077	22.0542	19.14882	17.7874	16.2918	15.3692
2	2474.4236	724.0906	705.5661	30.7088	29.7101	25.6274	19.4966	18.07732	16.3881	15.4932	14.7707
3	2288.6388	537.1949	456.1886	31.2498	26.3404	24.9594	19.7442	18.7194	17.9814	16.1636	15.4229
4	2257.5909	914.9179	757.4173	30.3188	26.6303	21.2085	21.0419	19.6400	16.4352	15.6005	14.7649
5	1397.4916	490.0064	471.9773	31.8041	27.1358	21.3741	19.4219	19.4209	16.5248	15.8792	15.4557
6	1386.1016	603.1896	518.9901	30.8450	26.3429	21.4612	19.3104	17.6198	17.2648	16.3788	14.7692
7	1024.6595	666.6348	664.5530	30.7786	26.2238	21.2123	19.5816	17.4790	16.3602	16.3551	14.8109
8	925.0678	462.3369	446.1551	30.9299	26.2436	21.2664	19.3921	17.4870	16.3125	15.2054	15.1587
9	925.0058	659.1381	658.4553	30.3014	26.5172	21.2429	19.3784	17.4848	16.3536	15.2046	14.7564
10	582.0920	441.5511	440.6515	30.2731	26.2238	21.2024	19.2785	17.4776	16.3122	15.1941	14.7355

According to the minimum values in Tables 5.2 and 5.3, the estimated values of p and q are 2 and 3 respectively, which is correct obviously. However, it should be noted that the estimation will be influenced by N, the model orders, and the noise. Usually, the larger the value of N and the higher the model orders, the more accurate the estimation. For example, the estimation of an ARMA(3,2) may be less accurate than that of an ARMA(6,4). If there is a noise in the sequence which is called as the observed noise usually, the accuracy of the estimation may be low.

Table 5.2. The derived values of the minimum eigenvalues.

p q	0	1	2	3	4	5	6	7	8	9	10
1	0.9963	1.4900	1.1742	0.4347	0.9881	1.1060	1.1326	1.0765	1.0849	1.0582	1.0396
2*	0.6434	0.9338	1.1973	0.3531*	1.1035	1.0543	0.8840	0.9440	0.9213	0.9510	0.9611
3	0.9249	0.74189	0.6466	1.0176	0.8866	0.9739	1.0127	1.0355	1.0972	1.0433	1.0442
4	0.9864	1.7031	1.6603	0.9702	1.0110	0.8497	1.0657	1.0492	0.9140	0.9652	0.9573
5	0.6190	0.5356	0.6231	1.0490	1.0190	1.0078	0.9230	0.9888	1.0055	1.0179	1.0468
6	0.9918	1.2310	1.0996	0.9698	0.9708	1.0041	0.9943	0.9073	1.0448	1.0315	0.9556
7	0.7392	1.1052	1.2805	0.9978	0.9955	0.9884	1.0140	0.9920	0.9476	0.9986	1.0028
8	0.9028	0.6935	0.6714	1.0049	1.0008	1.0026	0.9903	1.0005	0.9971	0.9297	1.0235
9	0.9999	1.4257	1.4758	0.9797	1.0104	0.9989	0.9993	0.9999	1.0025	0.9999	0.9735
10	0.6293	0.6699	0.6692	0.9991	0.9889	0.9981	0.9948	0.9996	0.9975	0.9993	0.9986

Table 5.3. The other derived values of the minimum eigenvalues.

p q	1	2	3*	4	5	6	7	8	9	10
0	0.1348	0.9645	0.3987	0.1362	0.8066	0.8860	0.9135	0.9217	0.9390	0.9603
1	0.2016	0.7600	0.1476	0.3096	0.9028	0.9073	0.8683	0.9289	0.9159	0.9434
2	0.2926	0.9744	0.0435	0.9675	0.8626	0.7608	0.9272	0.9066	0.9454	0.9534
3	0.2347	0.8492	0.0685	0.8429	0.9476	0.7911	0.9481	0.9606	0.8989	0.9542
4	0.4053	0.8279	0.0400*	0.8783	0.7964	0.9921	0.9334	0.8368	0.9492	0.9464
5	0.3506	0.9632	0.0674	0.8532	0.7877	0.9087	0.9999	0.8509	0.9609	0.9733
6	0.4352	0.8604	0.0594	0.8540	0.8147	0.8998	0.9125	0.9799	0.9487	0.9017
7	0.6506	0.9969	0.0463	0.8520	0.8089	0.9231	0.8926	0.9360	0.9997	0.9056
8	0.4998	0.9650	0.0693	0.8485	0.8103	0.9119	0.9018	0.9328	0.9321	0.9969
9	0.7126	0.9990	0.0460	0.8751	0.8011	0.9122	0.9023	0.9353	0.9297	0.9705
10	0.75867	0.9980	0.0687	0.8662	0.8085	0.9093	0.9066	0.9333	0.9315	0.9698

5.4.3.4. Modelling with the Observed Noise

In the above, there is no noise in the observation of x, but there is usually a noise between x and its observation y [9-14]. That is

$$y(k) = x(k) + v(k), \qquad (5.37)$$

where $v(k)$ is the white noise with zero mean and variance of σ_v^2 and uncorrelated with $w(k)$. There is

$$R_{yy}(\tau) = \begin{cases} R_{xx}(\tau) + \sigma_v^2 & \tau = 0 \\ R_{xx}(\tau) & \tau \neq 0 \end{cases} \tag{5.38}$$

Substitute (5.38) into (5.17) and yield

$$\begin{bmatrix} R_{yy}(q+1) & \cdots & R_{yy}(0) - \sigma_v^2 & \cdots & R_{yy}(q+1-p) \\ R_{yy}(q+2) & \cdots & R_{yy}(1) & \cdots & R_{yy}(q+2-p) \\ \vdots & \vdots & \vdots & \vdots & \vdots \\ R_{yy}(p) & \cdots & R_{yy}(p-q-1) & \cdots & R_{yy}(0) - \sigma_v^2 \\ R_{yy}(p+1) & \cdots & R_{yy}(p-q) & \cdots & R_{yy}(1) \\ \vdots & \vdots & \vdots & \vdots & \vdots \\ R_{yy}(p+s) & \cdots & R_{yy}(p+s-q-1) & \cdots & R_{yy}(s) \end{bmatrix} \begin{bmatrix} 1 \\ \phi_1 \\ \phi_2 \\ \vdots \\ \phi_p \end{bmatrix} = \mathbf{0},$$

$$\tag{5.39}$$

where s governs the number of the equations in (5.39) and $s \geq q$. (5.39) is called as the noise-compensated modified Yule-Walker (NCMYW) equations [9-10]. The AR parameters can be estimated by the following two approaches. The first is based on the so-called extended modified Yule-Walker (EMYW) equations. The EMYW equations are composed of the equations from the order of $p+1$ to the order of $p+s$ in (5.39) and $s \geq p$. Hence, the EMYW equations are the NCMYW equations excluding the former $p-q$ equations in (5.39). The EMYW equations can be solved by the least squares algorithm to estimate the AR parameters. The second method solves the NCMYW equations directly by the generalized eigenvalue problem approach to estimate the AR parameters and the observed noise variance σ_v^2 simultaneously. Since the lower order equations are included in the NCMYW equations, the AR parameter estimation accuracy of the second method is slightly higher than that of the first method but the accuracy difference between two methods is negligible [3].

With the assumption that $p > q$, the $p-q$ equations in (5.39) (from $q+1$ to p) are extracted from (5.39). It yields

$$\mathbf{R}_{yy}\boldsymbol{\theta} = \sigma_v^2 \mathbf{a}, \tag{5.40}$$

where $\mathbf{a} = \begin{bmatrix} \phi_{q+1} & \cdots & \phi_p \end{bmatrix}^{\mathrm{T}}$; and

$$\mathbf{R}_{yy} = \begin{bmatrix} R_{yy}(q+1) & R_{yy}(q) & \cdots & R_{yy}(0) & \cdots & R_{yy}(p-q-1) \\ R_{yy}(q+2) & R_{yy}(q+1) & \cdots & R_{yy}(1) & \cdots & R_{yy}(p-q-2) \\ \vdots & \vdots & \vdots & \vdots & \vdots & \vdots \\ R_{yy}(p) & R_{yy}(p-1) & \cdots & R_{yy}(p-q-1) & \cdots & R_{yy}(0) \end{bmatrix}$$

If the AR parameters have been estimated by the EMYW equations or the NCMYW equations, the observed noise variance σ_v^2 can be estimated as

$$\hat{\sigma}_v^2 = \left(\hat{\mathbf{a}}\hat{\mathbf{a}}^T\right)^{-1}\hat{\mathbf{a}}^T\mathbf{R}_y\hat{\boldsymbol{\theta}}, \tag{5.41}$$

where $\hat{\mathbf{a}}$ and $\hat{\boldsymbol{\theta}}$ are the estimation of $\mathbf{a}$ and $\boldsymbol{\theta}$ respectively. In (5.41), the estimation accuracy of σ_v^2 is determined by $\hat{\boldsymbol{\theta}}$ and $\mathbf{R}_y$. Since only the lower order equations in (5.39) are used to estimate σ_v^2, the lower lags of the ACF are included in $\mathbf{R}_y$, which can help to improve the estimation accuracy of σ_v^2. It is also noticed that the estimation accuracy of the AR parameters is relatively high. Hence, it can be expected of high estimation accuracy of σ_v^2 [3].

5.5. Allan Variance

5.5.1. Methodology

In 1966, D Allan proposed a method based on the variance analysis to analyze the stability of an atomic oscillator [15]. The method is called as Allan variance which became a standard to evaluate the performance of an atomic oscillator later. In 1983, M Tehrani introduced the method into the modelling for the random errors of a ring laser gyroscope (RLG) [16]. In 1998, the method became the IEEE standard to model the random errors of RLG. In 2003, the method was used to model the random errors of the micro-electro-mechanical system (MEMS) inertial sensors [17]. Now, Allan variance is one of the most popular methods to model the random errors of the inertial sensors [1]. In the following, the method will be introduced by taking a gyroscope as an example, but it is also suitable for an accelerometer or other sensors.

As illustrated in Fig. 5.6, the acquired random error samples of a gyroscope, $\{\delta\omega_k\}(k=1,2,\cdots,N)$, are grouped firstly. It is required that $n < N/2$. The sample period is T_s. The group number is K and $K = [N/n]$ where $[\cdot]$ is rounding.

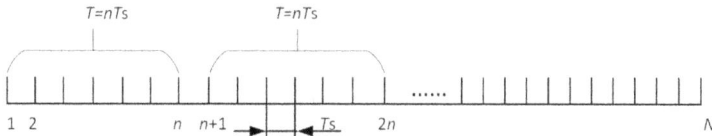

Fig. 5.6. The grouped samples for Allan variance analysis.

The mean of each group is

$$\delta\overline{\omega}_i(T) = \frac{1}{n} \sum_{j=(i-1)n+1}^{in} \delta\omega_j(T) \tag{5.42}$$

The Allan variance can be derived as

$$\sigma^2(T) = \frac{1}{2(K-1)} \sum_{k=1}^{K-1} \left[\delta\overline{\omega}_{k+1}(T) - \delta\overline{\omega}_k(T) \right]^2 \tag{5.43}$$

When the samples in a group, n, are changing, the Allan variance changing with nT can be obtained. The random errors can be estimated by the changing Allan variance, which will be introduced as follows.

Assume that the PSD of the ergodic and stationary process $\delta\omega$ is $\Phi_{\delta\omega}(f)$. According to (5.43), there is

$$\sigma^2(T) = 4\int_0^\infty \Phi_{\delta\omega}(f) \frac{\sin^4(\pi fT)}{(\pi fT)^2} df \tag{5.44}$$

The derivation of (5.44) is as follows. Since the process is assumed as ergodic, (5.43) can be written as

$$\sigma^2(T) = \frac{1}{2}\left\langle \left[\delta\overline{\omega}_{k+1}(T) - \delta\overline{\omega}_k(T) \right]^2 \right\rangle$$
$$= \frac{1}{2}\left\langle \delta\overline{\omega}_k^2(T) \right\rangle + \frac{1}{2}\left\langle \delta\overline{\omega}_{k+1}^2(T) \right\rangle - \left\langle \delta\overline{\omega}_k(T)\delta\overline{\omega}_{k+1}(T) \right\rangle, \tag{5.45}$$

where $\langle \cdot \rangle$ is the expectation by the samples. Since

$$
\begin{aligned}
\left\langle \delta\overline{\omega}_k^2(T) \right\rangle &= \frac{1}{T^2} \int_{t_k}^{t_k+T} \int_{t_k}^{t_k+T} \left\langle \delta\omega(t)\delta\omega(\alpha) \right\rangle \mathrm{d}\alpha \mathrm{d}t \\
&= \frac{1}{T^2} \int_{t_k}^{t_k+T} \int_{t_k}^{t_k+T} R_{\delta\omega}(t-\alpha) \mathrm{d}\alpha \mathrm{d}t \\
&= \frac{1}{T^2} \int_{-\infty}^{+\infty} \int_{t_k}^{t_k+T} \int_{t_k}^{t_k+T} \Phi_{\delta\omega}(f) e^{j2\pi f(t-\alpha)} \mathrm{d}\alpha \mathrm{d}t \mathrm{d}f \\
&= \int_{-\infty}^{+\infty} \Phi_{\delta\omega}(f) \frac{\sin^2(\pi f T)}{(\pi f T)^2} \mathrm{d}f
\end{aligned}
\tag{5.46}
$$

Moreover

$$
R_{\delta\omega}(t-\alpha) = \left\langle \delta\omega(t)\delta\omega(\alpha) \right\rangle
\tag{5.47}
$$

$$
\int_{t_k}^{t_k+T} \int_{t_k}^{t_k+T} e^{j2\pi f(t-\alpha)} \mathrm{d}\alpha \mathrm{d}t = \frac{\sin^2(\pi f T)}{(\pi f)^2}
\tag{5.48}
$$

In (5.46), the expectation is derived in the continuous form rather than in the discrete form in (5.42) for convenience. Since the process is assumed as stationary too, there is

$$
\left\langle \delta\overline{\omega}_{k+1}^2(T) \right\rangle = \left\langle \delta\overline{\omega}_k^2(T) \right\rangle
\tag{5.49}
$$

Similarly

$$
\begin{aligned}
\left\langle \delta\overline{\omega}_k(T)\delta\overline{\omega}_{k+1}(T) \right\rangle &= \frac{1}{T^2} \int_{t_k}^{t_k+T} \int_{t_k+T}^{t_k+2T} \left\langle \delta\omega(t)\delta\omega(\alpha) \right\rangle \mathrm{d}\alpha \mathrm{d}t \\
&= \frac{1}{T^2} \int_{-\infty}^{+\infty} \int_{t_k}^{t_k+T} \int_{t_k+T}^{t_k+2T} \Phi_{\delta\omega}(f) e^{j2\pi f(t-\alpha)} \mathrm{d}\alpha \mathrm{d}t \mathrm{d}f \\
&= \int_{-\infty}^{+\infty} \Phi_{\delta\omega}(f) e^{j2\pi fT} \frac{\sin^2(\pi f T)}{(\pi f T)^2} \mathrm{d}f
\end{aligned}
$$

(5.50)

Substitute (5.46), (5.49) and (5.50) into (5.45) and yield

$$\sigma^2(T) = \int_{-\infty}^{+\infty} \Phi_{\delta\omega}(f)\left(1 - e^{j2\pi fT}\right)\frac{\sin^2(\pi fT)}{(\pi fT)^2}\,df$$

$$= 2\int_{-\infty}^{+\infty} \Phi_{\delta\omega}(f)\frac{\sin^4(\pi fT)}{(\pi fT)^2}\,df - j\int_{-\infty}^{+\infty} \Phi_{\delta\omega}(f)\frac{\sin^2(\pi fT)\sin(2\pi fT)}{(\pi fT)^2}\,df$$

$$(5.51)$$

According to (5.4), the PSD of an ergodic process is even, so (5.51) can be simplified as (5.44).

In (5.44), the Allan variance can be derived if the PSD of a process is known. On the contrary, the PSD of the process can also be derived to model the process further if the Allan variance is known, which is the idea of the modelling for the random errors by the Allan variance. In the modelling, the process is usually modelled as the combination of several independent models. In the following, several typical models constructed by the Allan variance are introduced.

5.5.2. Quantization Noise

The noise due to digital quantization in sampling is called as quantization noise which can be equivalent to the output of a rectangular window with the input of a white noise. Hence, its PSD is

$$\Phi_{\delta\omega}(f) = (2\pi f)^2 T_s Q_z^2 \frac{\sin^2(\pi fT_s)}{(\pi fT_s)^2}, \qquad (5.52)$$

where Q_z is the quantization noise strength. If f is large enough, (5.52) can be approximated as

$$\Phi_{\delta\omega}(f) \approx (2\pi f)^2 T_s Q_z^2 \qquad (5.53)$$

Substitute (5.53) into (5.44) and performing the integration yield

$$\sigma_Q^2(T) = \frac{3Q_z^2}{T^2} \qquad (5.54)$$

That is

$$\sigma_Q(T) = \sqrt{3}\frac{Q_z}{T} \qquad (5.55)$$

201

In the log-log frame, the slope of the Allan variance is -1 for the quantization noise as shown in Fig. 5.7. In Fig. 5.7, the Allan variance decreases with the increase of T, so the Allan variance of the quantization noise will be large when T is small.

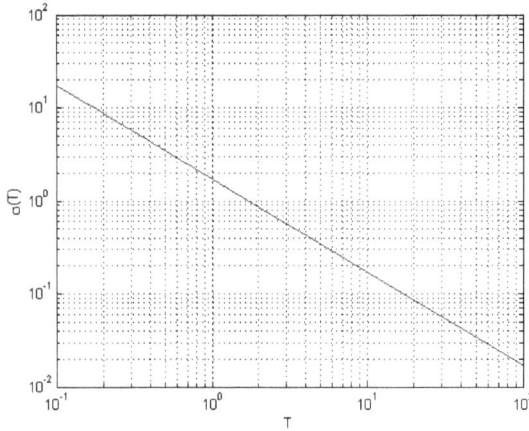

Fig. 5.7. Log-log plot of the Allan variance of the quantization noise.

5.5.3. Angular Random Walk

Angular random walk can be modelled as the white noise in the angular rate output of a gyroscope, so its PSD is

$$\Phi_{\delta\omega}(f) = Q^2, \tag{5.56}$$

where Q is the angular random walk coefficient. Similarly, substitute (5.56) into (5.44) and performing the integration yield,

$$\sigma_{ARW}^2(T) = 4\int_0^\infty Q^2 \frac{\sin^4(\pi fT)}{(\pi fT)^2}df = \frac{4Q^2}{\pi T}\int_0^\infty \frac{\sin^4 u}{u^2}du = \frac{Q^2}{T} \tag{5.57}$$

or

$$\sigma_{ARW}(T) = \frac{Q}{\sqrt{T}} \tag{5.58}$$

Fig. 5.8 depicts the log-log plot of the Allan variance of the angular random walk. The slope in Fig. 5.8 is -1/2.

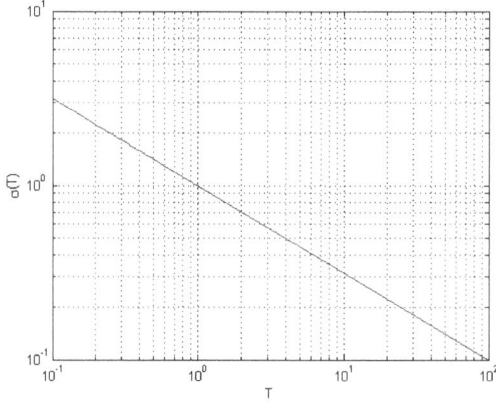

Fig. 5.8. Log-log plot of the Allan variance of the angular random walk.

5.5.4. Bias Instability

The PSD of a bias instability is

$$\Phi_{\delta\omega}(f) = \begin{cases} \dfrac{B^2}{2\pi f}, & f \le f_0 \\ 0, & f > f_0 \end{cases}, \tag{5.59}$$

where B is the bias stability coefficient and f_0 is the cut-off frequency. Substitute (5.59) into (5.44) and performing the integration yield

$$\sigma_B^2(T) = \frac{2B^2}{\pi}\int_0^{\pi Tf_0}\frac{\sin^4 u}{u^3}du = \frac{2B^2}{\pi}\Big\{\ln 2 -$$

$$-\frac{\sin^3(\pi Tf_0)}{2(\pi Tf_0)^2}\Big[\sin(\pi Tf_0)+4\pi Tf_0\cos(\pi Tf_0)\Big]+C(2\pi Tf_0)-C(4\pi Tf_0)\Big\}, \tag{5.60}$$

where

$$C(x) = \int_x^\infty \frac{\cos\alpha}{\alpha}d\alpha = \ln x + \sum_{k=1}^\infty (-1)^k \frac{x^{2k}}{2k(2k)!}+c, \tag{5.61}$$

203

where c is the Euler constant. According to (5.60), there is

$$\sigma_B(T) \rightarrow \begin{cases} 0, & T \ll \dfrac{1}{f_0} \\[2ex] \sqrt{\dfrac{2\ln 2}{\pi}}B, & T \gg \dfrac{1}{f_0} \end{cases} \tag{5.62}$$

Fig. 5.9 shows the log-log plot of the Allan variance of the bias instability. The slope in Fig. 5.9 is +1 when the correlation time is small but approaches 0 when it is large.

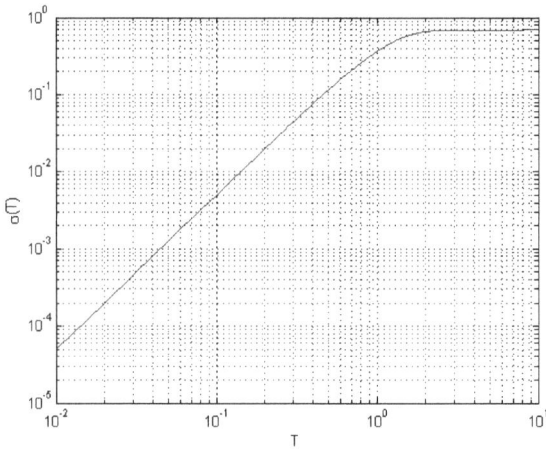

Fig. 5.9. Log-log plot of the Allan variance of the bias instability.

5.5.5. Rate Random Walk

The correlation time of a rate random walk is usually very large. The PSD of the rate random walk is

$$\Phi_{\delta\omega}(f) = \left(\frac{K}{2\pi f}\right)^2, \tag{5.63}$$

where K is the rate random walk coefficient. Substitute (5.63) into (5.44) and performing the integration yield

$$\sigma_{RRW}^2 (T) = \frac{K^2}{3} T \tag{5.64}$$

or

$$\sigma_{RRW} (T) = \frac{K}{\sqrt{3}} \sqrt{T} \tag{5.65}$$

Fig. 5.10 depicts the log-log plot of the Allan variance of the rate random walk. The slope in Fig. 5.10 is +1/2.

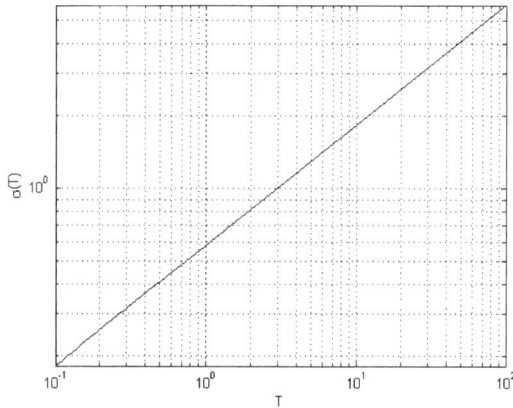

Fig. 5.10. Log-log plot of the Allan variance of the rate random walk.

5.5.6. Drift Rate Ramp

A drift rate ramp is deterministic. It is

$$\delta\omega = Rt, \tag{5.66}$$

where R is the drift rate ramp coefficient. Its PSD is

$$\Phi_{\delta\omega} (f) = \frac{R^2}{(2\pi f)^3} \tag{5.67}$$

Its Allan variance can also be derived by substituting (5.67) into (5.44), it will be acquired by substituting (5.66) into (5.42) and (5.43) directly.

$$\sigma_R^2(T) = \frac{R^2}{2}T^2 \tag{5.68}$$

or

$$\sigma_R(T) = \frac{R}{\sqrt{2}}T \tag{5.69}$$

Fig. 5.11 depicts the log-log plot of the Allan variance of the drift rate ramp and the slope is +1.

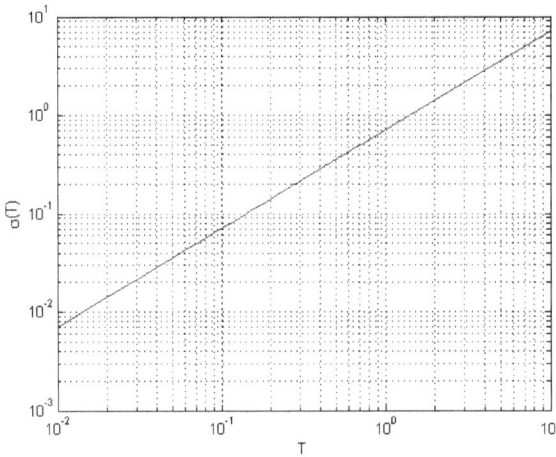

Fig. 5.11. Log-log plot of the Allan variance of the drift rate ramp.

5.5.7. First-Order Markov Process

The PSD of the first-order Markov process is

$$\Phi_{\delta\omega}(f) = \frac{(q_c T_c)^2}{1 + (2\pi f T_c)^2}, \tag{5.70}$$

where q_c is the driving noise strength and T_c is the correlation time. Substitute (5.70) into (5.44)and performing the integration yield

$$\sigma_M^2(T) = \frac{(q_c T_c)^2}{T}\left[1 - \frac{T_c}{2T}\left(3 - 4e^{-T/T_c} + e^{-2T/T_c}\right)\right] \tag{5.71}$$

$$\sigma_M^2(T) \rightarrow \begin{cases} \dfrac{(q_c T_c)^2}{T}, & T \gg T_c \\ \dfrac{1}{3} q_c^2 T, & T \ll T_c \end{cases} \qquad (5.72)$$

That is

$$\sigma_M(T) = \frac{q_c T_c}{\sqrt{T}} \sqrt{1 - \frac{T_c}{2T} \left(3 - 4e^{-T/T_c} + e^{-2T/T_c} \right)} \qquad (5.73)$$

Fig. 5.12 depicts the log-log plot of the Allan variance of the first-order Markov process. The slope in Fig. 5.12 approaches -1/2 when the correlation time is small and +1/2 when the correlation time is large.

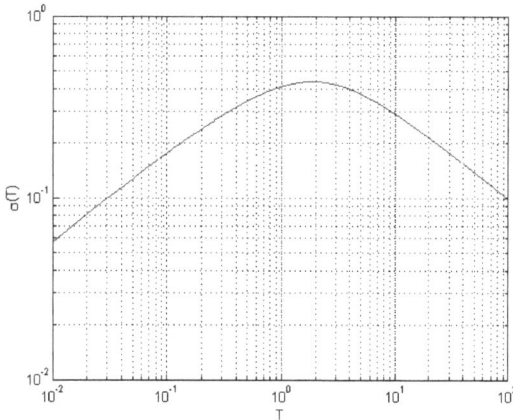

Fig. 5.12. Log-log plot of the Allan variance of the first-order Markov process.

5.5.8. Sinusoidal Noise

The PSD of a sinusoidal noise is

$$\Phi_{\delta\omega}(f) = \frac{1}{2} \Omega_0^2 \left[\delta(f - f_0) + \delta(f + f_0) \right], \qquad (5.74)$$

where Ω_0 is the amplitude and f_0 is the frequency. Substitute (5.74) into (5.44) and performing the integration yield

$$\sigma_S^2(T) = \Omega_0^2 \left[\frac{\sin^2(\pi f_0 T)}{\pi f_0 T} \right]^2 \tag{5.75}$$

or

$$\sigma_S(T) = \Omega_0 \frac{\sin^2(\pi f_0 T)}{\pi f_0 T} \tag{5.76}$$

Fig. 5.13 depicts the log-log plot of the Allan variance of the sinusoidal noise. The slope in Fig. 5.13 approaches +1 when the correlation time is small and -1 when the correlation time is large.

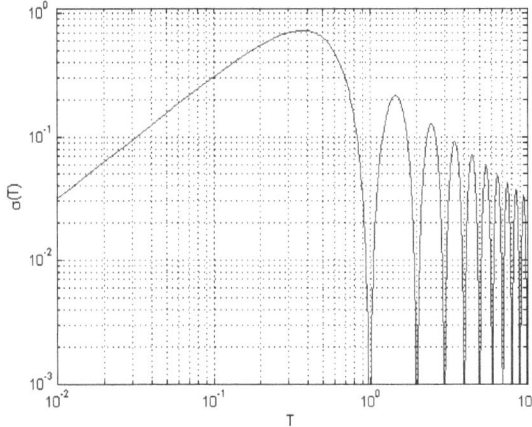

Fig. 5.13. Log-log plot of the Allan variance of the sinusoidal noise.

In applications, the real noise is usually composed of several above ones independently, so its Allan variance is the summation of those of the individuals as

$$\sigma^2(T) = \sigma_Q^2(T) + \sigma_{ARW}^2(T) + \sigma_{RRW}^2(T) + \\ + \sigma_R^2(T) + \sigma_B^2(T) + \sigma_M^2(T) + \sigma_S^2(T) \tag{5.77}$$

Hence, the Allan variance of the real noise can be fitted according to (5.77) in modelling, but the first-order Markov process and the sinusoidal noise are usually excluded. That is

$$\sigma^2(T) = \sigma_Q^2(T) + \sigma_{ARW}^2(T) + \sigma_B^2(T) + \sigma_{RRW}^2(T) + \sigma_R^2(T) =$$

$$= \frac{3Q_z^2}{T^2} + \frac{Q^2}{T} + \frac{2\ln 2}{\pi}B^2 + \frac{K^2}{3}T + \frac{R^2}{2}T^2 = \sum_{i=-2}^{2} A_i T^i, \qquad (5.78)$$

where A_i is the coefficient. Fig. 5.14 illustrates the log-log plot of the Allan variance of the real noise. The correlation time of one noise is different from that of another's in Fig. 5.14, so the noises can be modelled in the different ranges of the correlation time.

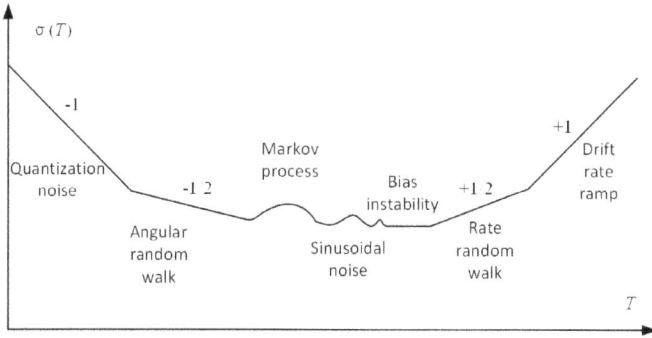

Fig. 5.14. Illustration of the Allan variance of the real noise.

Moreover, the accuracy of the Allan variance calculation varies with the group number. The calculating error of the Allan variance is

$$\delta\sigma(T) = \frac{\sigma(T,K) - \sigma(T)}{\sigma(T)}, \qquad (5.79)$$

where $\sigma(T)$ is the Allan variance without error, K is the group number, and N is the total samples. The standard deviation of $\delta\sigma(T)$ is derived as

$$\sigma_{\delta\sigma} = \frac{1}{\sqrt{2\left(\dfrac{N}{n} - 1\right)}} \qquad (5.80)$$

If the threshold of $\delta\sigma(T)$ is σ_{th}, the samples in a group should be

$$n < \frac{N}{\dfrac{1}{2\sigma_{\text{th}}^2} + 1} \tag{5.81}$$

Example 7

The errors of a gyroscope are depicted in Fig. 5.15. They can be modelled by the Allan variance as follows.

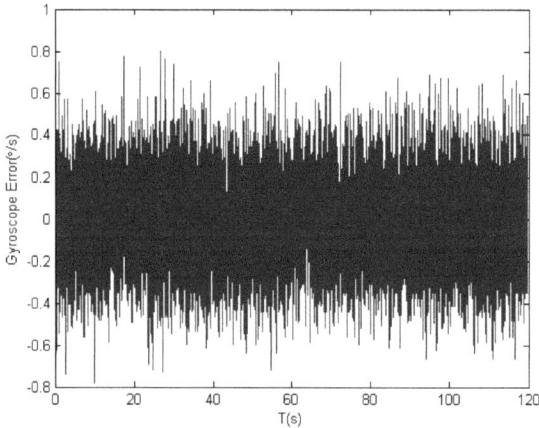

Fig. 5.15. The gyroscope noise.

Firstly, the samples in a group will be determined. Since the samples in a group should be less than half of the total samples at least, they should be less than 5985 as the total samples are 11971. The other aspect which should be taken into account when the samples in a group is determined is the accuracy of the calculated Allan variance in (5.81). Herein, let σ_{th} 10 % so that the samples in a group should be no more than 233. That is, K is in the range of [2, 233].

Then, calculate the Allan variances with K. The correlation time changes from 0.02 s to 2.33 s since the sample period is 0.01 s.

Next, fit the Allan variance of the noise to derive the coefficients of the individual noises. Since there are quantization noise, angular random walk, and bias instability possibly according to the Allan variance of the noise, the least squares fitting result is

$$\begin{cases} Q_z = 2.26'' \\ Q = 0.02°/s^{1/2} \\ B = 0.015°/s \end{cases}$$

Finally, the fitted Allan variance can be compared with the original one in Fig. 5.16, which proves the effectiveness of the modelling to some extent.

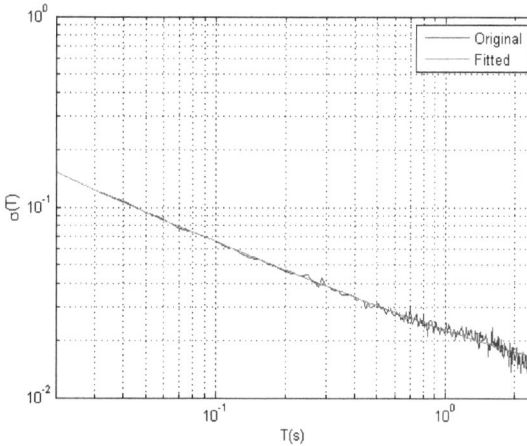

Fig. 5.16. The fitted and the original Allan variances.

5.6. Conclusions

Although the colored noise can be modelled by the different methods, such as the abovementioned three ones, the accuracies of the models should be similar if the modelling is effective. What makes the difference is the scope of the methods. The shaping filter is very simple and convenient to implement, but the scenarios that it can be used are limited. The ARMA modelling is very adaptive to the broad implementations, but its result is usually lacks the conceptual meaning, which is not helpful for its understanding. The modelling results of the Allan variance are provided with the exact meanings so that the method has been widely employed by manufacturers. However, if there are too many the colored noises modelled by the Allan variance to approach the original noise, it is no good for the following error compensation. For example, the augmented states will be increased significantly if there are too many the

211

colored noises in a Kalman filter. Moreover, the methods provided herein are for the steady errors of the inertial sensors or others', but they may lose the effectiveness more or less if the errors are time variant. To model the time-variant errors effectively, the improved methods such as the on-line modelling based on the on-line calibration have to be explored, which will not discuss herein.

References

[1]. N. El-Sheimy, H. Y. Hou, X. J. Niu, Analysis and modeling of inertial sensors using Allan variance, *IEEE Trans. Instrum. Meas.*, Vol. 57, 2008, pp. 140-148.

[2]. S. M. Seong, J. G. Lee, C. G. Park, Equivalent ARMA model representation for RLG random errors, *IEEE Trans. Aerosp. Electron. Syst.*, Vol. 36, 2000, pp. 286-290.

[3]. K. D. Wang, Y. X. Wu, Y. F. Gao, Y. Li, New methods to estimate the observed noise variance for an ARMA model, *Measurement*, Vol. 99, 2017, pp. 164-170.

[4]. L. H. Brandenburg, H. E. Meadows, Shaping filter representation of nonstationary colored noise, *IEEE Inf. Theo.*, Vol. 17, 1971, pp. 26-31.

[5]. A. Nehorai, P. Stoica, Adaptive algorithms for constrained ARMA signals in the presence of noise, *IEEE Trans. Acou. Speech Sig. Pro.*, Vol. 36, 1988, pp. 1282-1291.

[6]. S. Li, B. W. Dickinson, Application of the lattice filter to robust estimation of AR and ARMA models, *IEEE Trans. Acou. Speech Sig. Pro.*, Vol. 36, 1988, pp. 502-512.

[7]. C. Ran, Z. Deng, Information fusion multi-stage identification method for multisensory multi-channel ARMA models, in *Proceedings of the International Conference on Mechatronics, Electronics and Automotive Engineering (ICMEAE'11)*, 2011, pp. 2216-2221.

[8]. G. Liang, D.M. Wilkes, J. A. Cadzow, ARMA model order estimation based on the eigenvalues of the covariance matrix, *IEEE Trans. Sig. Pro.*, Vol. 41, 1993, pp. 3003-3009.

[9]. S. A. Fattah, W. P. Zhu, M. O. Ahmad, Identification of autoregressive moving average systems based on noise compensation in the correlation domain, *IET Sig. Pro.*, Vol. 5, 2011, pp. 292-305.

[10]. C. E. Davila, A subspace approach to estimation of autoregressive parameters from noisy measurements, *IEEE Trans. Sig. Pro.*, Vol. 46, 1998, pp. 531-534.

[11]. W. X. Zheng, Autoregressive parameter estimation from noisy data, *IEEE Trans. Circ. Syst. II Anal. Dig. Sig. Pro.*, Vol. 47, 2000, pp. 71-75.

[12]. W. X. Zheng, Fast identification of autoregressive signals from noisy observations, *IEEE Trans. Circ. Syst. II Anal. Dig. Sig. Pro.*, Vol. 52, 2005, pp. 43-48.

[13]. W. X. Zheng, On estimation of autoregressive signals in the presence of noise, *IEEE Trans. Circ. Syst. II Exp. Brief*, Vol. 53, 2006, pp. 1471-1475.

[14]. A. Mahmoudi, M. Karimi, Inverse filtering based method for estimation of noisy autoregressive signals, *Sig. Pro.*, Vol. 91, 2011, pp. 1659-1664.

[15]. D. Allan, Statistics of atomic frequency standards, *Proceedings of IEEE*, Vol. 54, 1966, pp. 221-230.

[16]. M. Tehrani, Ring laser gyro data analysis with cluster sampling technique, *Proceedings of SPIE*, Vol. 412, 1983, pp. 207-220.

[17]. S. Y. Bae, K. J. Hayworth, K. Shcheglov, D. V. Wiberg, High performance MEMS micro-gyroscope, *Proceedings of SPIE*, Vol. 4755, 2002, pp. 316-324.

Chapter 6

Open Channel Cross Section Design: Review of Recent Developments

Said M. Easa

6.1. Introduction

Open-channel flow is a flow that is not entirely included within rigid boundaries. The design of open channel involves two aspects: channel alignment and cross section. Channel alignment is generally controlled by topography, right-of-way, and adjacent structures/transportation facilities. The design of channel cross section (or simply section) involves four elements (Fig. 6.1): section shape, boundary type, longitudinal slope, and design criteria. Based on a review of 117 publications, this chapter focuses on recent developments of channel section design with respect to all elements, except longitudinal slope which is beyond the scope of this review.

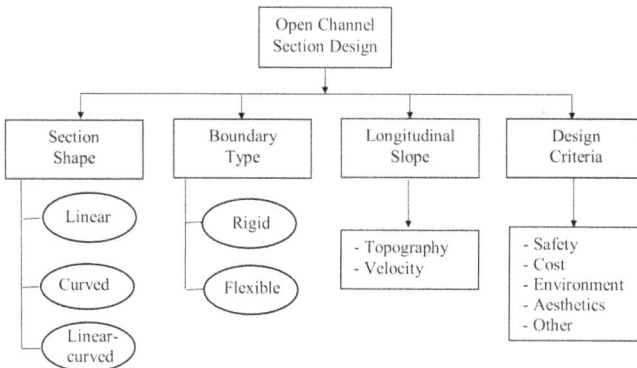

Fig. 6.1. Elements of open channel section design.

Said M. Easa
Department of Civil Engineering, Ryerson University, Toronto, Ontario, Canada

Section shape may be linear (e.g. trapezoidal), curved (e.g. parabolic), or linear-curved (e.g. parabolic-bottom triangle). Conventional sections have been mostly linear or curved, but recently there has been a surge of new linear-curved sections that have better performance. The types of section boundaries are classified as rigid and flexible, see Kilgore and Cotton [1]. Rigid boundaries are used to reduce seepage, increase discharge capacity, and prevent erosion. The boundary is normally a manufactured lining, such as cast-in-place concrete. Flexible boundaries may be permanent (e.g. riprap) or temporary (e.g. bare soil). The design method of the channel section depends on the boundary type. Rigid sections are normally designed to minimize construction cost or maximize hydraulic efficiency, while flexible sections are designed to minimize erosion of the channel using such methods as permissible tractive force and permissible velocity.

The longitudinal slope of the channel is influenced by the terrain through which the channel alignment is planned. An important factor in the selection of the longitudinal slope is water velocity which should be within the stated criteria. The longitudinal slope is normally used in section design as input. However, some analysis of the velocity, especially at the planning stage, would be needed to determine the acceptable longitudinal slope. The process of determining longitudinal slopes is addressed in hydraulic design manuals [2].

Once the section shape, boundary type, and longitudinal slope are determined, the dimensions of the section are determined considering several criteria, including hydraulics, cost, safety, environment, and aesthetics. The majority of design methods are based on hydraulics and cost. In addition, some methods consider the uncertainties of the input variables. In practice, a number of alternative sections should be selected and evaluated to determine the best section that satisfies specified criteria [2]. Various methods for determining optimal section dimensions are presented in this chapter. Open channel design has been addressed to varying degrees in a number of textbooks, including Akan [3], Chaudhry [4], French [5], and Chow [6].

This chapter is organized as follows. Section 6.2 presents a brief discussion of the historical developments of section shapes. Section 6.3 reviews conventional open channel sections (1894-2009). Section 6.4 describes five new sections that have inspired further developments (2003-2009). Section 6.5 describes seven subsequently developed channel sections (2010-2018). Section 6.6 describes three design methods for determining section dimensions, including best hydraulic (most efficient) section, most economic section, and probabilistic methods. Section 6.7 describes developments related to the design of flexible channels. Section 6.8 discusses several design considerations. Section 6.9 presents concluding remarks.

6.2. Historical Development

The historical development of channel sections is divided into four periods that overlap, as shown in Table 6.1. The first period (1894-2009) is characterized by the development and implementation of conventional sections, such as trapezoidal, rectangular, parabolic, power-law (PL), and horseshoe. The second period (2003-2009) witnessed the development of five innovative sections (linear and linear-curved): polygonal section, trapezoidal-rectangular section, section with parabolic sides and horizontal-bottom (HB), trapezoidal section with round corners, and section with two-segment parabolic sides and HB. These new sections have inspired further developments.

During the third period (2010-2018), more new sections that have provided better performance or more flexibility were developed. The sections included the following seven shapes: two-segment linear sides with HB, multi-segment linear sides with HB, ice-covered trapezoidal section, ordinary elliptic sides with HB, general elliptic sides with HB, cubic parabolic sides with HB, and super-best hydraulic PL section. The super-best section represents a new interesting discovery of an optimal section of the general PL section.

In the final period (1991-2017), new topics have been addressed in the design of channel sections. These included incorporating uncertainty in the design of conventional and ice-covered trapezoidal sections, development of iterative and direct methods for the design of flexible channels, such as grass-lined channels and channels lined with riprap, cobble, or gravel. Other topics included explicit solutions for different types of flow depths (e.g. normal and critical), selection of section freeboard (vertical length from water surface to channel top), consideration of seepage in channel design, and development of a new polynomial section with smooth top corners, which is especially useful for the design of compound channels.

Table 6.1. Historical developments of various shapes
of open channel sections.

Category	Section Family	Section Type [a]	Reference	Year
Conventional	Linear	Trapezoidal Rectangular Triangular Compound	Chow [6]	1959
	Curved	Horseshoe	Carson et al. [40]	1894
	Curved	Circular Parabolic	Chow [6]	1959
	Curved	Power-law	Strelkoff and Clemmens [36]	2000
	Linear-curved	Pipe-handle	Carson et al. [40]	1894
	Linear-curved	P-bottom triangle	Babaeyan et al. [43]	2000
	Linear-curved	C-bottom triangle	Chahar and Basu [44]	2009
Inspiring Initiatives	Linear	Polygonal	Kurbanov and Khanov [45]	2003
	Linear	Trapezoidal-rectangular	Abdulrahman [48]	2007
	Linear-curved	PS-HB	Das [49]	2007
	Linear-curved	Trapezoidal-RC	Froehlich [53]	2008
	Linear-curved	TSPS-HB	Easa [54, 56]	2009
Recent Developments	Linear	TSLS-HB	Vatankhah [55]	2010
	Linear	MSLS-HB	Easa [58]	2011
	Linear	Trapezoidal-IC	Han et al. [59]	2017
	Linear-curved	Ordinary ES-HB	Easa and Vatankhah [60]	2014
	Linear-curved	General ES-HB	Easa [61]	2016
	Linear-curved	Cubic parabolic sides-HB	Han and Easa [63]	2017
	Curved	Super-best power-law	Han and Easa [62]	2018
New Topics	Linear (probabilistic)	Trapezoidal	Cesare [67]	1991
	Linear (probabilistic)	Trapezoidal	Easa [68-70]	1992
	Linear (probabilistic)	Trapezoidal	Das [71-73]	2007
	Linear (probabilistic)	Trapezoidal	Adarsh et al. [74-76]	2012
	Linear (probabilistic)	Trapezoidal-IC	Easa [77]	2017
	Linear (direct)	Trapezoidal-flexible-grass	Easa and Vatankhah [87]	2012
	Linear (direct)	Trapezoidal-flexible-RCG	Easa et al. [79]	2015

[a] P = parabolic, C = circular, PS = parabolic sides, HB = horizontal bottom, TSPS = two-segment parabolic sides, RC = round corners, TSLS = two-segment linear sides, MSLS = multi-segment linear sides, IC = ice-covered, ES = elliptic sides, and RCG = riprap, cobble, or gravel.

6.3. Conventional Sections (1894-2009)

Conventional sections are classified here into three families (Fig. 6.2): linear family, curved family, and linear-curved family. The linear family includes sections consisting of only linear segments, while the curved family includes sections consisting of only curved segments. The linear-curved family includes sections with both linear and curved segments. This classification aids the discussion of subsequent developments to improve the conventional sections.

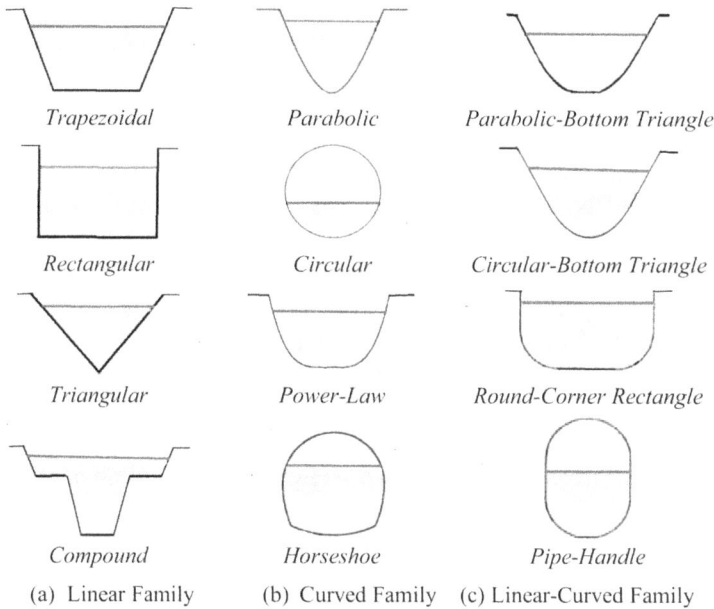

Trapezoidal   *Parabolic*   *Parabolic-Bottom Triangle*

Rectangular   *Circular*   *Circular-Bottom Triangle*

Triangular   *Power-Law*   *Round-Corner Rectangle*

Compound   *Horseshoe*   *Pipe-Handle*

(a) Linear Family   (b) Curved Family   (c) Linear-Curved Family

Fig. 6.2. Conventional families of channel sections based on shape.

6.3.1. Linear Family

The linear family includes trapezoidal, rectangular, triangular, and compound (Fig. 6.2a). The characteristics of the trapezoidal section and its special cases (rectangular and triangular) are well documented in the literature [3-6]. The optimal design of these sections has been addressed in numerous studies, including Swamee and Bhatia [7], Guo and Hughes [8], Das [9], Jain et al. [10], Aksoy and Altan-Sakarya [11], and Han et al. [12]. In addition, Froehlich [13] evaluated the best hydraulic trapezoidal section when the width and depth are constrained. Among

219

the linear family, compound sections present some challenges and interesting research has been conducted to overcome them.

A compound section is composed of several distinct subsections, each has its own roughness coefficient. When the flood plains are wide and shallow compared to the main channel, prediction of the normal discharge of the compound section will be challenging. In the conventional method, the compound section is simply divided into three subsections and then the subsection discharges are added. This method is not accurate. The reason is that as the flow starts to cover the flood plain, section flow area slowly increases, while wetted perimeter rapidly increases. Thus, the hydraulic radius, velocity, and discharge decrease with the increase in flow depth, which is not logical [5].

To overcome the challenge of estimating the discharge in such compound sections, considerable research has been conducted. The accuracy of different ways of defining the subsections has been evaluated by Posey [14]. Modified, more accurate methods have also been developed until about 2000 by Shiono and Knight [15], Wark et al. [16], Ackers [17], Wormleaton and Merrett [18], Lambert and Myers [19], Bousmar and Zech [20], and Ervine et al. [21]. To overcome the extensive computations required by the preceding methods, more general models have been subsequently developed (1997-2010) using regression analysis by Hosseini [22] and artificial neural networks by MacLeod [23], Liu and James [24], Zahiri and Dehghani [25], and Unal et al. [26].

More recently, Azamathulla and Zahiri [27] developed a precise dimensionless model for predicting the discharge for a compound channel using linear genetic optimization. The model was calibrated using published stage–discharge data for 394 laboratories and field data for 30 compound channels. The discharge ratio in a compound section was expressed as a function of dimensionless parameters, as follows

$$\frac{Q_t}{Q_b} = f\left(D_r, COH, \frac{Q_{DCM}}{Q_b} \right), \tag{6.1}$$

where Q_t is the total discharge, Q_b is the bankfull discharge, D_r is the depth ratio (ratio of flow depth in the floodplain to that of the main channel), COH is the coherence parameter, and Q_{DCM} is the discharge calculated using Manning formula assuming vertical dividing planes between the main channel and the floodplains. The three dimensionless parameters of Eq. (6.1) have been used by other researchers [17, 19, 28, 29].

6.3.2. Curved Family

The curved family includes parabolic, circular, power-law, and horseshoe (Fig. 6.2b). The most popular in this family is the parabolic section. The optimal characteristics of this section have been addressed by numerous authors, including Mironenko et al. [30], Loganathan [31], Chahar [32], Chahar et al. [33], and Han et al. [12]. Compared with trapezoidal sections, the advantages of parabolic channels have been discussed in the literature and include strength, ease of construction, cost, speed, and aesthetics. Mironenko et al. [30] and Laycock [34] pointed out the following specific advantages: (a) the shell shape of parabolic channels offers greater structural strength with no weak points that might occur in a trapezoidal channel, (b) parabolic channels eliminate the problems of slumping or tension cracking, which normally occurs near the base of trapezoidal slopes during concrete placement, (c) parabolic channels are more economical than the equivalent trapezoidal channels, and (d) the method of construction of parabolic channels is faster, especially when slip-forming is used.

The mathematical form of the power-law section is generic and makes it possible to model natural channels and some specific well-known man-made open channel shapes, such as rectangular, trapezoidal, triangular, and parabolic. This versatility is the main attraction of this type of section, as pointed out by Hussein [35] who also presented a concise review of previous developments of the PL section. Despite its practical attractiveness, the main difficulty of designing a PL section is the calculation of the wetted perimeter. Based on the work of Strelkoff and Clemmens [36], Anwar and de Vries [37] developed two approximate expressions for the relative wetted perimeter for specific exponent values, ignoring the freeboard. The two approximations had a discontinuity at an exponent equal to 0.5 and this discontinuity was resolved later by Kacimov [38]. Considering a PL section with freeboard, Anwar and Clarke [39] estimated the relative wetted perimeter using the same approach as that of [36] and obtained the best hydraulic section for a given maximum side slope and relative freeboard using Lagrange multiplier method. In 2008, Hussein [35] used the isoperimetric theory and inequality to accurately approximate the wetted perimeter of PL section with freeboard using a single expression.

The circular and horseshoe sections are commonly used in the sewer systems. Their developments date back to the 19th century, see Carson et al. [40]. The characteristics of the standard horseshoe can be found in Merkley [41]. A comprehensive review of the studies related to the performance of sewer sections can be found in Hager [42]. Another curved section, called catenary has been described in some textbooks [6]. However, the optimal characteristics of this section have not been explored in the literature.

6.3.3. Linear-Curved Family

The linear-curved family includes parabolic-bottom triangle (PBT), circular-bottom triangle (CBT), round-corner rectangle, and pipe-handle sections (semi-circular soffits with vertical walls), as shown in Fig. 6.2c. The pipe-handle section, and a variety of other shapes such as egg-shaped and gothic section with a pointed arch are used in the sewer systems [42].

Limited research has been conducted for the sections of this linear-curved family. Babaeyan-Koopaei et al. [43] developed the optimal design of the PBT section and its characteristics. The best hydraulic section was determined using Lagrange multiplier method. The section parameters were compared with those of the trapezoidal, parabolic, and CBT sections. The results showed that for all values of side slopes, the flow area and wetted perimeter of the PBT section were less than those of the trapezoidal and parabolic sections for the same discharge. This finding implies that the PBT section was more economical than the trapezoidal and parabolic sections. The PBT section was slightly less economical than the CBT section, as expected.

Chahar and Basu [44] developed optimal design equations for the CBT and PBT sections. The authors determined the best hydraulic section, subject to the Manning formula, using Lagrange multiplier method. In addition, an optimization model for the most economic section (based on earthwork and lining costs) was presented and solved using *Solver* of MS Excel. The optimization model included constraints on section variables, such as top width, side slop, and flow depth. The proposed design equations were in explicit form and produced optimal channel dimensions in a single step.

6.4. New Inspiring Initiatives (2003-2009)

During the period 2003-2009, a few authors started to geometrically modify the conventional shapes and developed new sections to improve performance. These inspiring new sections provided new ideas and appeared to be the basis for subsequent developments. The sections are grouped into two categories: inspiring linear sections and inspiring linear-curved sections.

6.4.1. Inspiring Linear Sections

Polygonal Section

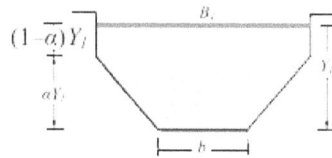

The semi-circular section is known to be the best hydraulic channel section. However, construction of such a section is technically an arduous task and is expensive. For this reason, in 2003 Kurbanov and Khanov [45] proposed a polygonal section that provided a close approximation to the semi-circular section, as shown in Fig. 6.3a. The best hydraulic polygonal section was developed. This polygonal section was based on the work of Uginchus [46]. In another paper [47], the authors developed the critical depth for two cases: two-segment linear sides (Fig. 6.3a) and three-segment linear sides.

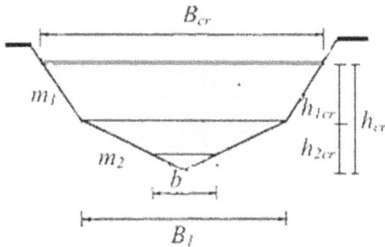

(a) Polygonal section [45]   (b) Trapezoidal-rectangular section [48]

Fig. 6.3. Geometry of inspiring linear sections.

Trapezoidal-Rectangular Section

In 2007, Abdulrahman [48] proposed a section that consisted of a trapezoidal section with a rectangular section at the top, as shown in

223

Fig. 6.3b. The characteristics of this section, including wetted perimeter and flow area were presented and the best hydraulic section was developed. The results showed that the best hydraulic section is a half-octagon and the proposed section is more efficient than the trapezoidal section.

6.4.2. Inspiring Linear-Curved Sections

Section with Parabolic Sides and HB

The first initiative to improve the performance of the conventional parabolic sections was made in 2007 by Das [49]. The author proposed a section with a horizontal bottom and parabolic sides (HBPS), as shown in Fig. 6.4a. The HB and sides have different roughness coefficients and the composite roughness was calculated based on Horton's formula [50, 51]. The feasibility of the new section which was lined as a composite channel was evaluated. The optimal design that minimized construction cost subject to constraints on channel capacity and other imposed restrictions on section dimensions was developed using Lagrange multiplier method.

The flow area A and top width T_f of the HBPS section are given by

$$A = y \left\{ b + \frac{4}{3} (z_1 + z_2) \sqrt{y + f} \sqrt{y} \right\}$$ (6.2)

$$T_f = b + 2(z_1 + z_2)(y + f)$$ (6.3)

The wetted perimeter can also be calculated in a closed-form, see Federal Highway Administration [52]. Other section characteristics can be found in Das [49].

In model evaluation, various design scenarios including unrestricted design and design with constraints on flow depth, side slope, and top width were considered. Each of the design scenarios was evaluated considering fixed and depth-dependent freeboards. The optimization results showed that the proposed section provided cost saving compared with the trapezoidal section.

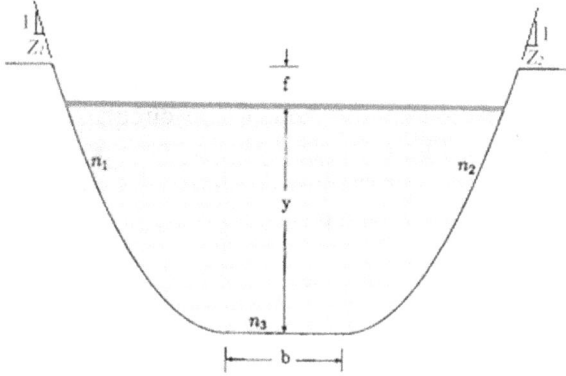

(a) Section with parabolic sides and HB [49]

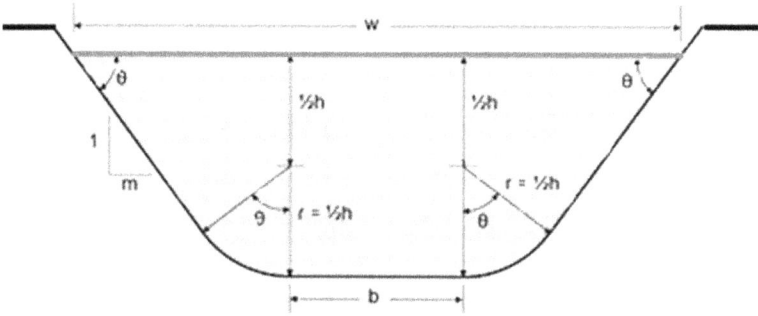

(b) Trapezoidal section with round corners (shallow depth) [53]

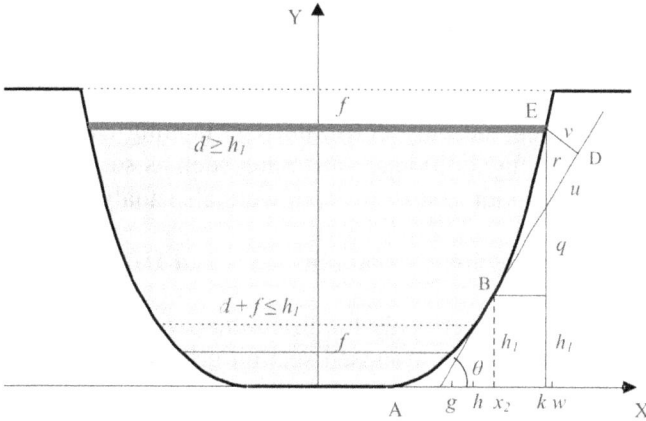

(c) Section with two-segment parabolic sides and HB [54]

Fig. 6.4. Geometry of inspiring new linear-curved sections.

Trapezoidal Section with Round Corners

In 2008, Froehlich [53] proposed a generalized trapezoidal shape in which a horizontal bottom was added to a circular-bottom triangular section, as shown in Fig. 6.4b. This figure corresponds to the case of deep excavation depth. For shallow depth, r is set equal to h. The flow area A, wetted perimeter P, and top width w are given by

$$A = (b + \alpha h)h \tag{6.4}$$

$$P = b + \beta h \tag{6.5}$$

$$w = b + \delta h \tag{6.6}$$

$$\alpha = \alpha(\kappa, m) = \left[1 - 2\kappa(1 - \kappa)\right]m + 2\kappa(1 - \kappa)\sqrt{1 + m^2} + \kappa^2 \arctan \frac{1}{m} \tag{6.7}$$

$$\beta = \beta(\kappa, m) = 2\left[\kappa m + (1 - \kappa)\sqrt{1 + m^2} + \kappa \arctan \frac{1}{m}\right] \tag{6.8}$$

$$\delta = \delta(\kappa, m) = 2\left[(1 - \kappa)m + \kappa\sqrt{1 + m^2}\right] \tag{6.9}$$

Setting $k = 1$ gives the section with $r = h$, while the section of Fig. 6.4b results when $k = 0.5$. When $k = 0$, the expressions for A and P are reduced to those of the conventional trapezoidal section.

This general trapezoidal section is reduced to a trapezoidal section when $r = 0$ and to a circular-bottom triangular section when $b = 0$. The author developed optimal section dimensions using Lagrange multiplier method for conditions when the only constraint is Manning-based discharge. Other cases involving additional constraints such as section side slope, bottom width, top width, and flow depth were also analyzed.

Section with Two-Segment Parabolic Sides and HB

In 2009, Ease [54] geometrically modified the parabolic section side to improve performance. The proposed section consisted of two-segment parabolic sides (TSPS) and a horizontal bottom, as shown in Fig. 6.4c. Since this section is somewhat more complex than linear segments, its characteristics are presented here in more detail.

The height of the first parabolic segment, y, for specified x is given by

$$y = h_1 (x - x_1)^2 / b_1^2, \quad x_2 \geq x \geq x_1 \tag{6.10}$$

The first and second derivatives of y are given by

$$y' = 2h_1(x - x_1)/b_1^2 \tag{6.11}$$

$$y'' = 2h_1/b_1^2 \tag{6.12}$$

The second parabolic segment has two tangents, one is the common tangent at B and the other is the tangent at C. Then, the height of this segment, Y, is given by

$$Y = h_1[(x_3 - x)/b_2]^2 + S_2(x - x_2), \quad x \geq x_2, \tag{6.13}$$

where S_2 is the slope of the second tangent which is given by

$$S_2 = h_2 / b_2 \tag{6.14}$$

As noted, for $x = x_2$, $Y = h_1$ and for $x = x_3$, $Y = h_2$. The first and second derivatives of Y of Eq. (6.13) are given by

$$Y' = -2h_1[(x_3 - x)/b_2^2] + S_2 \tag{6.15}$$

$$Y'' = 2h_1/b_2^2 \tag{6.16}$$

To ensure that the two parabolic segments have a common tangent at B, the following condition for h_2 was developed

$$h_2 = 2h_1 (b_1 + b_2) / b_1 \tag{6.17}$$

Note that Point C is not a physical point of the section side, but rather is just a point used to define the parabola of the second segment. The length of the second segment that is used in the actual channel section could be smaller or larger than BC, depending on the flow depth and the freeboard. Based on the preceding characteristics, the section area of flow and wetted perimeter were developed [54].

Compared with the HBPS section of Das [49], the TSPS section provided less construction cost for the same input conditions. In addition, it was shown that for certain constraints on top width and side slope, the solution of the HBPS section was infeasible, while the TSPS section provided an optimal solution. The reason is that the TSPS section has more degrees of freedom to handle such constraints.

6.5. Following Developments (2010-2018)

Since the developments of the five inspiring channel sections presented in the previous section, several new channel sections that offered different capabilities were developed. The new sections are grouped into three categories: (1) linear family, (2) elliptic family, and (3) power-law family.

6.5.1. Linear Family

Section with Two-Segment Linear-Sides

In 2010, Vatankhah [55], in a discussion of the TSPS section by Easa [54], proposed a section with two-segment linear sides (TSLS), as shown in Fig. 6.5a. The section was more economical than the trapezoidal section. It was shown, however, that the TSLS section may provide invalid results since the side's lower segment may become part of the horizontal bottom due to lack of constraints [56]. In 2014, Vatankhah [57] introduced two types of semi-regular polygons (Types I and II) as the best hydraulic sections. Type I is a semi-regular polygon with flat bottom and Type II is a semi-regular polygon with angled bottom, similar to Kurbanov and Khanov [45].

Section with Multiple-Segment Linear Sides

In 2011, Easa [58] developed a new section with multiple piece-wise linear sides (Fig. 6.5b). The flow depth d is the sum of the linear segment heights y_i and flow area A and wetted perimeter P_s are given by

$$A = bd + \sum_{i=1}^{m}\left(2dz_i y_i - z_i y_i^2\right) - 2\sum_{i=1}^{m-1}(\sum_{j=1}^{i}y_i)z_{i+1}y_{i+1} \qquad (6.18)$$

$$P_s = \sum_{i=1}^{m}y_i\left(1+z_i^2\right)^{0.5}(z_i < z_s) \qquad (6.19)$$

The new section was found to be more economical than the TSLS and TSPS sections. With many segments, the section side becomes smooth.

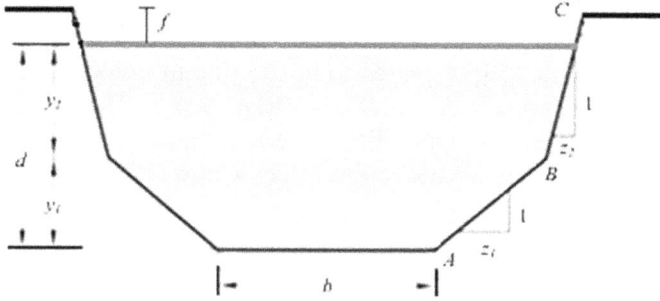

(a) Section with two-segment linear sides and horizontal bottom [55]

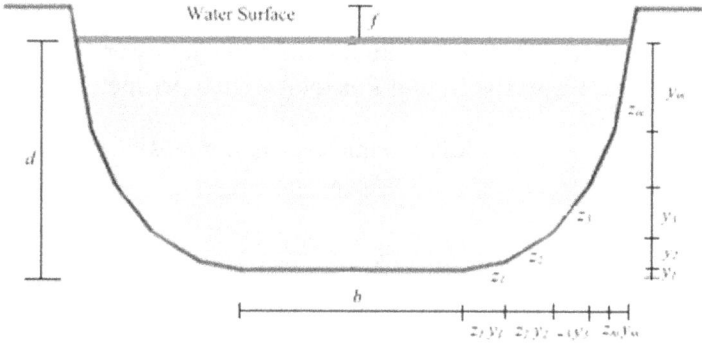

(b) Section with multiple-segment linear sides and horizontal bottom [58]

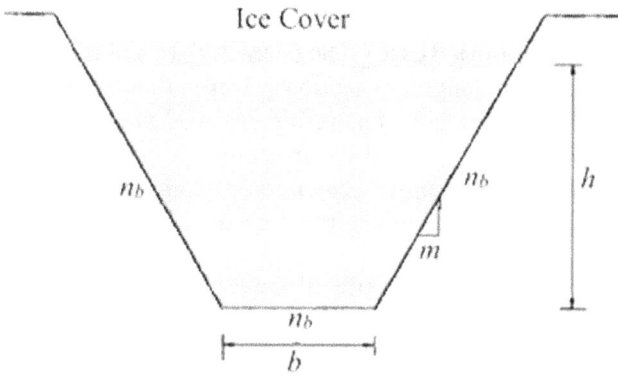

(c) Geometry of ice-covered trapezoidal section [59].

Fig. 6.5. Recent sections of multiple linear-segment family.

Ice-Covered Trapezoidal Section

In 2017, Han et al. [59] appeared to be the first to introduce the design of ice-covered trapezoidal sections (Fig. 6.5c). The authors first developed the general optimality condition based on the following general formula of composite roughness by Chow [6]

$$n_t = \frac{PR^{5/3}}{\left(\dfrac{R_b^{5/3} P_b}{n_b} + \dfrac{R_i^{5/3} P_i}{n_i}\right)} , \qquad (6.20)$$

where n_t is the composite roughness coefficient, n_b and n_i are the roughness coefficients for the channel bed (channel bottom and sides) and ice cover, P_b and P_i are the wetted perimeter of the channel bed and ice cover, R_b and R_i are the hydraulic radii of the lower bed zone and upper ice zone, respectively, and P is total wetted perimeter ($P_b + P_i$).

Then, using a dimensionless parameter, the composite roughness was derived as

$$n_t = \frac{\left(P_i \epsilon^{3/2} n_i^{3/2} + P_b n_b^{3/2}\right)^{5/3}}{\left(P_i + P_b\right)^{3/2} \left(P_i \epsilon^{5/2} n_i^{3/2} + P_b n_b^{3/2}\right)} , \qquad (6.21)$$

where ϵ is a dimensionless parameter (V_i / V_b), where V_b and V_i are the average velocities of the lower bed zone and the upper ice zone.

The best hydraulic section was developed for two cases: $\epsilon \neq 1$ (general case) and $\epsilon = 1$ (simplified case). The characteristics of the best hydraulic trapezoidal and rectangular sections were then developed using Lagrange multiplier method. The results showed that the best hydraulic sections under ice-covered and ice-free conditions were different. The best hydraulic section under ice-covered conditions was relatively narrow and deep, which would help reduce the heat loss at water surface.

What is interesting is that the optimal section of the ice-covered section, which is intended for transporting water during the winter season, can also be used to transport water in the ice-free seasons. The reason for this is that the wetted perimeter (discharge) of the ice-covered section is larger (less) than that of the ice-free section. Therefore, a channel designed for a certain discharge under ice-covered conditions can accommodate a larger discharge in the warm seasons.

6.5.2. Elliptic Family

Ordinary Elliptic Section

In 2014, Easa and Vatankhah [60] proposed a section with elliptic sides and HB (Fig. 6.6a). The elliptic side is mathematically represented by

$$\frac{x^2}{B_0^2} + \frac{(y-Y_0)^2}{Y_0^2} = 1, \quad (Z_m > 0), \tag{6.22}$$

where (B_0, Y_0) and (B, Y) and are the semi axes of the ellipses for $Z_m > 0$ and $Z_m = 0$, respectively, B is the width of each elliptic side at the ground surface, and Y is the total depth of the channel $(d + f)$, where d is the flow depth and f is the freeboard.

The ordinary elliptic section produces, as special section types, a section with HB and circular sides, a circular section, and a rectangular section. Exact and approximate formulas for the area and perimeter of the section were derived. These formulas were then used to develop an optimization model for determining the optimal section dimensions that minimized total construction cost. In addition, the best hydraulic elliptic section was derived. The constraints of the model included discharge and physical requirements, such as flow depth, top width, and side slope with fixed or depth-dependent freeboard. The results showed that the new section was substantially more economical and more flexible than the PL section.

General Elliptic Section

In 2016, Easa [61] proposed a section with general elliptic (GE) sides and HB. The idea of this section came from the general ellipse which is represented by

$$\left|\frac{x}{a}\right|^m + \left|\frac{y}{b}\right|^n = 1, \tag{6.23}$$

where a and b are the semi axis of the general ellipse and m, n are shape parameters (> 0). A special case of the GE ellipse corresponds to $m = n$ which is the super-ellipse shown in Fig. 6.6b for n ranging from 0.3 to 50. The super-ellipse defines a closed curve contained in the rectangle $-a \leq x \leq a$ and $-b \leq y \leq b$, where a and b are the semi-diameters of the curve. For $0 < n < 1$, the super-ellipse is a four-armed star with concave (inward) sides. The special case $n = 2$ corresponds

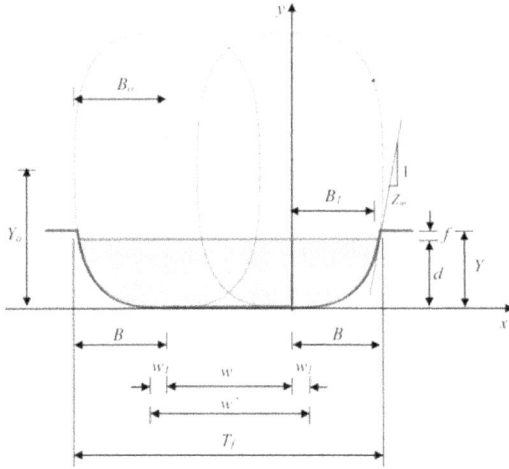

(a) Ordinary elliptic section [60].

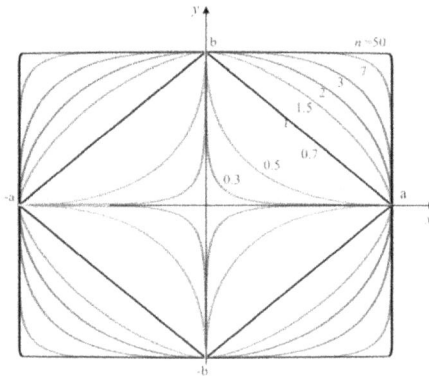

(b) General elliptic curves for $m = n$ [61].

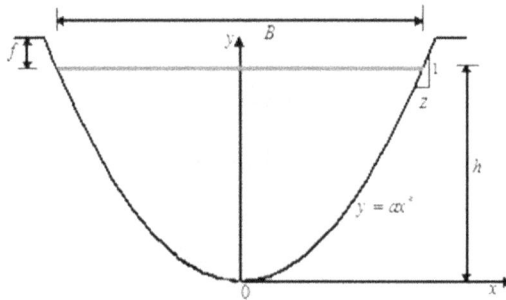

(c) General power-law section [62].

Fig. 6.6. Recent sections of elliptic and power-law families.

to the ordinary elliptic curve which has been implemented by Easa and Vatankhah [60]. Clearly, only the family of GE curves with m and $n \geq 1$ (integer and non-integer) are applicable to channel section design. The geometry of the GE section is similar to that of Fig. 6.6a, where the section has general instead of ordinary elliptic sides.

Considering a Cartesian coordinate axes with an origin at the right side of the horizontal bottom, where the x axis is horizontal, similar to Fig. 6.6a, the GE is described by

$$\left(\frac{x}{B_0}\right)^m + \left(\frac{y - Y_0}{Y_0}\right)^n = 1, \, (Z_m > 0), \tag{6.24}$$

where (B_0, Y_0) are the semi axes of the GE section. The section has dimensions B, Y, and w, where B is the width of the elliptic side at the bank level, Y is the total channel depth, and w is the bottom width. For $Z_m = 0$, $Y_0 = Y$ and $B_0 = B$, and the semi axes of the GE become (B, Y).

The GE section is versatile as it produces numerous special sections, including elliptic, parabolic, circular, triangular, trapezoidal, and rectangular. In addition, the section produces a wide variety of sections that correspond to many combinations of integer and non-integer values of the parameters m and n. Obviously, the GE section would be more economical than all of its special case sections.

6.5.3. Power-Law Family

Considering the symmetry of the PL section about its centerline (Fig. 6.6c), the section is defined using the following general form [62]

$$y = a|x|^k, \quad k \geq 1, \tag{6.25}$$

where y is the vertical coordinate (m), a is the shape factor, x is the horizontal coordinate (m), and k is the exponent. The shape of the PL section varies with k. For example, for $k = 1$ the section is triangular and for $k = 2$ the section is parabolic.

Cubic Section

In 2016, Han and Easa [63] explored the hydraulic performance of a cubic PL section ($k = 3$) and found that its performance was better than the parabolic section ($k = 2$). Subsequently, the authors explored section performance for another exponent ($k = 8/3$) and found that the performance of this section [64] was better than that of the cubic section. These findings motivated the authors to explore other sections with exponents $k = 4$ and 5. Surprisingly, the performance of these sections was worse than that for $k = 3$ and 8/3. Then, it was hypothesized that a super-performing PL section exists somewhere between $k = 3$ and 4. This has led to the development of the exact solution of the PL section considering k as a parameter and then the super-best hydraulic (SBH) section, as described next.

Super-Best Power-Law Section

Previous methods have developed the best hydraulic PL section for specific k. In 2018, Han and Easa [62] presented a general exact solution of the best hydraulic PL section with the exponent k as a parameter. The method was based on Gauss Hyper-geometric function and Lagrange multiplier method. The relationships between k and each of the optimum width-depth ratio and side slope were derived. An explicit formula of the wetted perimeter for a wider range of k was developed. Also, explicit formulas for flow depth, flow area, wetted perimeter, and shape factor as functions of k were developed using the least-squares fitting method.

Using the exact solution, the SBH section was developed. This section represents a new discovery as it provides the global maximum discharge among all possible PL section shapes [62]. The optimum exponent of this section is $k = 3.3471$ and its dimensionless parameter $\eta = B/h = 2.1278$, where B is the water surface width and h is the flow depth. The discharge Q, flow area A, wetted perimeter P, and shape factor a of the SBH section are given by

$$Q = 1.0416 \frac{h^{8/3}\sqrt{S}}{n} \tag{6.26}$$

$$A = 1.5890\left(Qn/\sqrt{S}\right)^{3/4} \tag{6.27}$$

$$P = 3.1826\left(Qn/\sqrt{S}\right)^{3/8} \tag{6.28}$$

$$a = 0.8425\left(Qn/\sqrt{S}\right)^{-0.8802} \tag{6.29}$$

The normal and critical depths of the SBH section are given by

$$h = 0.9848 \left(Qn / \sqrt{S} \right)^{3/8} \tag{6.30}$$

$$h_c = 0.8650 \frac{\sqrt[5]{\beta Q^2 g^4}}{g} \tag{6.31}$$

What is interesting is that the research showed that the area of the super-best hydraulic PL section is only slightly larger than that of the semi-circular section (a difference of 0.36%) which is known to provide the largest capacity under the same flow area among all section shapes. Besides, the super-best PL section will eliminate the construction difficulties associated with the vertical slope of the semi-circular section.

6.6. Design Methods

The design methods for channel sections are based on steady uniform flow. The most commonly used uniform-flow formula worldwide is the Manning formula which is simple and produces acceptable accuracy in most practical applications. Three design methods are presented in this section: best hydraulic section, most economic section, and probabilistic methods.

6.6.1. Best Hydraulic Section

The best hydraulic section is typically obtained by minimizing the wetted perimeter (or section area) subject to an equality constraint for the discharge, normally based on the Manning formula. The optimization model is usually solved using Lagrange multiplier method. To provide the reader with highlights of this method, the general solution by Monadjemi [65] is presented here.

Consider the case where the flow area, wetted perimeter, and discharge are expressed in terms of the following variables: b (bottom width), h (flow depth), and z (side slope, expressed as horizontal: vertical). The channel longitudinal slope S is given. The discharge of uniform flow can be expressed using the Manning formula as,

$$Q = \frac{A^{5/3} S^{1/2}}{n \, P^{2/3}}, \tag{6.32}$$

where Q is the flow discharge (m^3/s), A is the flow area (m^2), S is the longitudinal slope of channel bottom, n is the roughness coefficient, and P is the wetted perimeter (m). The best hydraulic section is obtained by solving the following optimization model which minimizes the flow area subject to the discharge constraint:

$$\text{Minimize } A = f(h, b, z) \tag{6.33}$$

Subject to

$$\Phi = Q - \frac{A^{5/3} S^{1/2}}{n\, P^{2/3}}, \tag{6.34}$$

where Φ is the constraint function.

Using Lagrange multiplier method to solve this problem, the necessary conditions for the best hydraulic section are expressed as

$$\frac{\partial A}{\partial b}\frac{\partial P}{\partial h} = \frac{\partial A}{\partial h}\frac{\partial P}{\partial b} \tag{6.35}$$

$$\frac{\partial A}{\partial b}\frac{\partial P}{\partial z} = \frac{\partial A}{\partial z}\frac{\partial P}{\partial b} \tag{6.36}$$

Solving Eqs. (6.35) and (6.36) yields the dimensions of the best hydraulic section. Monadjemi [65] applied this method to rectangular, trapezoidal, and round-bottom triangular sections. Numerous studies have explored the best hydraulic section for various section shapes [39, 43, 44, 48, 57, 59, 65].

6.6.2. Most Economic Section

Optimization methods to determine the most economic (least-cost) section have been developed. Application of these methods is important for lined channels to reduce total construction cost. A general solution method developed by Han and Easa [12] is used here for illustration.

The most economic section is a section that has the least construction cost of the channel, given the same conditions of discharge, roughness, and bottom slope. The total cost is composed mainly of excavation, lining, and land acquisition costs. To develop the general optimality condition, the objective function of the optimization model that considers safe freeboard a is written as

$$\text{Minimize } C = C_e A^* + C_l P^* + C_a B^*,$$ (6.37)

where C is the total construction cost per unit length of the channel ($), C_e is the excavation cost per unit area of the section ($/m^2$), A^* is the section area considering freeboard (m^2), C_l is the lining cost per unit length of the section ($/m), P^* is the lining length considering freeboard (m), C_a is the land acquisition cost per unit length of the top of the section ($/m), and B^* is the top width of the section considering freeboard (m). The optimization model of Eqs. (6.37) and (6.34) is nonlinear and a general direct method for its solution was developed using Lagrange multiplier method [12].

For the trapezoidal section, A, P, A^*, P^*, and B^* are determined in terms of h and b for a given z. For a general section, assume that the two variables (h and b) are denoted by x_1 and x_2. Then, the following equations are equivalent to the optimization model of Eqs. (6.37) and (6.34)

$$\frac{\partial C}{\partial x_1} + \lambda \frac{\partial \Phi}{\partial x_1} = 0$$ (6.38)

$$\frac{\partial C}{\partial x_2} + \lambda \frac{\partial \Phi}{\partial x_2} = 0,$$ (6.39)

where λ is the Lagrange multiplier. Eliminating λ from Eqs. (6.38) and (6.39) yields

$$\frac{\partial C}{\partial x_1} \frac{\partial \Phi}{\partial x_2} = \frac{\partial \Phi}{\partial x_1} \frac{\partial C}{\partial x_2}$$ (6.40)

This equation is applicable to any type of open channel section. Substituting the first partial derivatives of Φ with respect to x_1 and x_2 into Eq. (6.40) yields

$$5\frac{\partial A}{\partial x_2}\frac{\partial C}{\partial x_1}P + 2\frac{\partial C}{\partial x_2}\frac{\partial P}{\partial x_1}A = 5\frac{\partial A}{\partial x_1}\frac{\partial C}{\partial x_2}P + 2\frac{\partial C}{\partial x_1}\frac{\partial P}{\partial x_2}A$$ (6.41)

which is the general differential equation for the most economic section (trapezoidal, parabolic, or any other section). Solving Eq. (7.41) and the Manning formula of Eq. (6.34), the optimal dimensions of the most economic sections (h and b) can be obtained. Numerous studies have been conducted to determine the dimensions of the most economic sections [7, 8, 10, 12].

It is important to note that most models in the literature have adopted construction cost as the evaluation criterion. This cost includes two items (excavation and surface lining) that are directly affected by the dimensions of the channel section. Generally, construction cost of open channels involves other items, such as cost of land and equipment, engineering fees, overhead cost, and miscellaneous cost. However, these items are likely to be the same regardless of the type of channel section used. In addition, other aspects (e.g., operation and maintenance costs, seepage cost, flood damage potential, and environmental impacts) are considered in the preliminary design stage, where a lifecycle analysis is usually conducted to evaluate different alternatives involving channel location, shape, and materials [2].

6.6.3. Probabilistic Methods

Probabilistic methods have been developed for the design of open channel sections, but research in this area has been somewhat limited. The methods include reliability analysis and stochastic optimization. The early effort to incorporate reliability analysis into channel design appears to start in 1980 with the work of Tung and Mays [66] who developed static and dynamic risk and reliability models based on first-order reliability method (FORM). The models were used to develop risk-safety curves for culvert design. In 1991, Cesare [67] addressed the uncertainties associated with the peak channel discharge and other variables of open channel design.

In 1992, Easa [68] presented a method for designing open channels that incorporated the uncertainty associated with various design variables. The design involved two random components: runoff (demand) and channel capacity (supply). The runoff was formulated considering the uncertainties of rainfall intensity, drainage area, and other watershed characteristics. Channel capacity was formulated considering the uncertainties of roughness coefficient, longitudinal slope, channel width, side slope, and water-flow depth. The probabilistic characteristics of the runoff and channel capacity were established based on FORM. In 1994, Easa [69, 70] extended the reliability analysis to consider multiple failure modes. The first failure mode occurs when the runoff exceeds channel capacity. The second failure mode occurs when the actual flow velocity exceeds the maximum allowable velocity for erosion control. The third failure mode occurs when the actual velocity is less than the minimum allowable velocity for deposition. The failure probability of each mode was estimated using advanced FORM. System failure probability that accounted for the correlations between the failure modes was presented. The method was verified using Monte Carlo simulation.

In 2007, Das [71] developed a probabilistic optimization model for composite trapezoidal sections. The model included two objective functions that minimized total construction cost and flooding probability subject to uniform-flow equation as a constraint. The flooding probability constraint was developed using first-order analysis. The single-objective optimization model was solved using Lagrange multiplier method. In 2008, Das [72] developed a mathematical model for chance-constrained optimal design of composite trapezoidal sections. The model maximized the probability of channel capacity being greater than the design flow and minimized the probable construction cost. In 2010, Das [73] used the criteria of minimization of construction cost and flooding probability to verify the acceptability of the proposed HBPS section. The multi-objective optimization model was solved similarly to the trapezoidal section. In 2013, this work was further extended by Adarsh and Sahana [74] who used meta-heuristic optimization to solve the model.

In 2010, Reddy and Adarsh [75] evaluated the design of composite trapezoidal sections using stochastic search algorithms such as genetic algorithms, particle swarm optimization (PSO), and elitist mutated PSO considering parametric uncertainty. Constrained channel design was performed considering a comprehensive objective function involving a chance constrained model that simultaneously accounts for flow exceedance and overtopping probability.

In 2012, Adarsh [76] presented a modified comprehensive model for the design of open channel sections considering uncertainties in seepage loss, evaporation loss, and construction cost (land acquisition, lining, and excavation). The model was solved using a probabilistic global search Lausanne (PGSL). Parameter uncertainty was modeled using FORM. A multi-objective function that minimized construction cost and the probability of overtopping was used subject to chance-constrained overtopping probability and channel capacity. The proposed coupled FORM-PGSL approach offered a novel tool for the optimal design of rigid open channels.

In 2017, Easa [77] presented a probabilistic optimization model for ice-covered channels that incorporated the uncertainties of the roughness coefficients of both ice cover and channel bed. The model determined the best hydraulic trapezoidal section by minimizing section area subject to design discharge. The nonlinear discharge equation was linearized using Taylor series which was found to be very accurate. The proposed chance-constrained model represented a first step toward a comprehensive uncertainty model for ice-covered section design.

6.7. Design of Flexible Channels

Kilgore and Cotton [1] presented design procedures for the following types of flexible linings: (a) riprap, cobble, and gravel lining, (b) vegetative lining and bare soil, (c) manufactured lining, and (d) gabion lining. The design is based on the permissible tractive force method, see for example Raudkivi [78].

The method includes two parts. First, the flow conditions are computed based on channel geometry, design discharge, roughness, channel alignment, and channel slope. Second, the required erosion protection is determined by computing the shear stress on the channel lining at the design discharge and comparing that stress to the permissible value for the type of lining/soil that makes up the channel boundary. The method involves a trial procedure. In this section, highlights of the design method for the first two types of linings are presented.

6.7.1. Riprap, Cobble, and Gravel Linings

Iterative Design Method

The method involves three types of iterations related to specified design discharge, side slope stability, and channel geometry. Given the dimensions of the trapezoidal section (B and Z), channel slope S, and specified discharge Q_s, an acceptable mean stone size for channel bottom and sides is calculated. The method involves the following steps, as described by Kilgore and Cotton [1]:

Step 1. Select a trial (initial) $D50_a$, based on the available sizes for the project and determine its specific weight.

Step 2. For the first iteration, select an initial channel depth d_i,. For the subsequent iterations, estimate a new depth, as follows

$$d_{i+1} = d_i (Q/Q_i)^{0.4}, \qquad (6.42)$$

where Q_i is the discharge calculated in iteration i. Determine d_a and $d_a/D50_a$.

Step 3. Calculate roughness coefficient n and the discharge Q_i.

Step 4. If Q_i is within 5 % of Q_s, proceed to Step 5. If not, go back to Step 2 and estimate a new flow depth.

Step 5. Calculate the particle Reynolds number, and determine Froude number and safety factor. Calculate the required $D50_b$ for shear stress on channel bottom.

Step 6. If the calculated $D50_b$ is greater than the trial size in Step 1, then the trial size is unacceptable for design. Repeat Steps 2-5 with new trial value of $D50_a$. Otherwise, the previous trial size is acceptable. However, if the required mean stone size is much smaller than the previous trial size, the design procedure should be repeated at Step 2 with a smaller, more cost-effective mean stone size. As an alternative, B and/or Z may be modified to produce a lower depth of flow, and in turn satisfy the selected $D50_a$.

Step 7. Check side slope stability, given $D50_b$ for channel bottom (Step 6) by calculating K_1, K_2, and the stable $D50_s$ for the side slope. The suitability of the selected stone class size is then determined similarly to Step 6. If the available stone class sizes are not satisfactory, different values of channel width or side slope can be tried to satisfy the design criteria.

The trial method is generally easy to follow and simple to apply. However, if the lining type and channel dimensions should be modified to satisfy the design criteria (discharge, permissible shear stresses, and side slope stability), the method becomes extremely complex as many trial combinations would be required to determine a feasible solution. To aid design, a direct method has been recently developed.

Direct Design Methods

In 2015, Easa et al. [79] presented a direct method for the design of sections with riprap, cobble, or gravel linings using optimization. The optimization model directly provided the best channel lining subject to the specified design criteria. The user needs to specify available stone sizes and then the model selects the best size for straight and curved channel segments. The optimization model can be easily solved using *Solver* software. The proposed model provided efficiency in the design of roadside channels and would be useful in practice.

In 2016, Gupta et al. [80] presented a design method for a trapezoidal earthen channel, where the side slopes were riveted with riprap stones and the bottom was unlined. The performance of different types of stones (round, sub-round and sub-angular, and angular) was evaluated based on minimum cost and performance effectiveness using particle swarm optimization. The results showed that channels with angular riprap offered the minimum construction cost, followed by sub-round and sub-angular, and round stones. In addition, the freeboard impacted channel shape, dimensions, and cost. Other studies on stable channel sections include Bhattacharjya and Satish [81], Froehlich, [82], and Easa et al. [83].

6.7.2. Grass-Lined Channels

In grass-lined channels, the roughness coefficient varies with the hydraulic radius. Therefore, the conventional design is based on a trial procedure [1]. The trial procedure is necessary because the roughness coefficient is a highly non-linear function of the hydraulic radius. The design of these channels is also unique because the roughness coefficient varies with the condition of the grass cover. For this reason, the design of such channels is performed in two stages.

The first stage uses the lowest retardance class of the grass and focuses on the design for channel stability with immature vegetation that offers little protection. The second stage uses the highest retardance classes of the grass and focuses on the design for channel capacity with mature vegetation offering greater flow resistance. The side slope of grass-lined channels should not be steeper than a ratio of 2:1 to aid grass growth and to make maintenance easier. When the channel is planned to be crossed by large equipment, z should be equal to 8 or greater [84]. The hydraulic resistance characteristics of grassed-lined channels are discussed by

Escarameia et al. [85]. More details on the design of grassed-lined channels can be found in Kilgore and Cotton [1].

The trial procedure has been eliminated by Akan and Hager [86] who developed two design charts for determining bottom width and flow depth for $z = 2$ based on predetermined solutions to the Manning formula for grass-lined channels. The charts were useful for quick design of grass-lined channels. For larger values of z, the charts overestimated the required dimensions. Subsequently, Easa and Vatankhah [87] developed direct analytical and graphical solutions (exact and nearly exact) for any value of z within the practical range $2 \leq z \leq 8$. In addition, Vatankhah and Easa [88] developed a simplified accurate solution for the design of erodible trapezoidal channels.

Note that an elegant mathematical trick used for developing the preceding direct solutions was the use of dimensionless parameters. Such parameters have been used by many authors to generate direct solutions for complex systems. Dimensionless parameters can also make the developed solutions more applicable and valid in forecasting.

6.8. Design Considerations

6.8.1. Normal and Critical Depths

Explicit solutions for different types of flow depth have been developed since the governing equations for these depths are normally implicit. Such explicit solutions promote practical applications. In particular, two types of depths (normal and critical) received more attention by researchers. The normal depth occurs in a steady uniform flow and is used in channel design, operation, flow measurement, flood control, and maintenances [5, 6]. The critical depth, which classifies gradually varied flow into subcritical and supercritical, is important for analyzing water surface profile, hydraulic jump, and energy dissipation.

A summary of the explicit solutions developed by researchers is presented in Table 6.2. The table shows the reference, date, section shape, and depth type. As noted, the section shapes are trapezoidal, rectangular, triangular, circular, parabolic, egg-shaped, horseshoe, compound, semielliptical, city-gate, power-law, wide rectangular, elliptical, cubic parabola, and general power-law.

Table 6.2. Explicit formulas for different section types and flow depths.

Reference	Section Shape	Depth Type [a]	Year
Swamee [89]	Trapezoidal Rectangular Triangular Circular	N	1994
Swamee and Rathie [90, 91]	Trapezoidal Rectangular Circular	N, C	2004-2005
Zhao et al. [92]	Circular	N	2008
Achour and Khattaoui [93]	Parabolic	N, C	2008
Raikar et al. [94]	Egg-shaped	N, C	2010
Vatankhah and Easa [95]	Trapezoidal Circular Horseshoe	N, C	2011
Liu et al. [96]	Compound	C	2012
Vatankhah [97-100]	Trapezoidal Semielliptical City-gate parabolic Power-law	N, C	2012-2015
Swamee and Rathie [101, 102]	Parabolic Wide rectangular Triangular	N	2012-2016
Li and Gao [103]	Parabolic	N	2014
Liu and Wang [104]	City-gate	N, C	2013
Perez et al. [105]	Elliptical	N, C	2015
Elhakeem [106]	Trapezoidal	N, C	2017
Han et al. [59]	Trapezoidal-Ice covered	N, C	2017
Han and Easa [62, 63]	Cubic parabola General power law	N, C	2016-2018

[a] N = normal depth and C = critical depth.

The idea of developing explicit solutions of flow depth started in 1994 by Swamee [89] who developed normal depth formulas for trapezoidal, rectangular, triangular, and circular sections. Starting in 2004, explicit formulas for other section shapes (or for the same shape but with improved accuracy) have been subsequently developed by Swamee and Rathie [90, 91], Zhao et al. [92], Achour and Khattaoui [93], Raikar et al. [94], Vatankhah and Easa [95], Liu et al. [96], Vatankhah [97-100], Swamee and Rathie [101, 102], Li and Gao [103], Liu and Wang [104], Perez et al. [105], Elhakeem [106], Han et al. [59], and Han and Easa

[62, 63]. Most references focused on both normal and critical depths, with a few addressing only normal or critical depth. In addition, Elhakeem [106] developed explicit solutions for the alternate and sequent depths from the specific energy and force equations, respectively.

6.8.2. Freeboard

A suitable amount of freeboard should be provided to prevent water surface from overtopping the banks. The freeboard is the vertical distance between the water surface (for design flow at normal depth) and the top of channel banks or channel lining [5]. The freeboard accounts for the uncertainty in design, construction, and operation of the channel. It depends on a number of factors, such as size of channel, flow velocity, flow depth, alignment curvature, interception of storm runoff, fluctuations in water level due to channel operation, tidal action, and wind action [5, 6]. For flexible channels, the freeboard can be estimated using the formula suggested by the U.S. Bureau of Reclamation as [107]

$$f = \sqrt{Cy} , \qquad (6.43)$$

where f is the freeboard (m), C is the coefficient that depends on the discharge Q ($C = 0.8$ and 1.4 for $Q = 0.5$ m^3/s and 85 m^3/s, respectively), and y is the design depth (m). For rigid (lined) channels, the following formula has been used for estimating freeboard [49, 108]

$$f = a + by^c , \qquad (6.44)$$

where a, b, and c are the coefficients. As noted, $b = 0$ corresponds to fixed freeboard and $a = 0$ corresponds to depth-dependent freeboard. For fixed freeboard, the value of a depends on the discharge Q, such that $a = 0.1$-0.15 m, 0.3 m, 0.5 m, 0.6 m, and 0.75 m for $Q < 0.06$ m^3/s, 0.06-1.0 m^3/s, 1.0-1.5 m^3/s, 5-10 m^3/s, and > 10 m^3/s, respectively [33].

The depth-dependent freeboard is more general as it differentiates between shallow and deep channels. For this type of freeboard, the values recommended by Chow [108] are $b = 0.5$, and c ranges from 0.6735 for $Q = 0.57$ m^3/s to 0.8723 for $Q \geq 85$ m^3/s. However, for the depth-dependent freeboard b may also vary with the discharge. For this reason, the following formula for b has been proposed by Chahar et al. [33].

$$b = 0.6722 + 2.3569x10^{-3}Q, \qquad (6.45)$$

where Q is the discharge (m^3/s). Note that b is assumed to vary linearly with Q.

6.8.3. Seepage Loss

Consideration of seepage loss in channel section design is important. Several researchers have modeled seepage loss in design. Kacimov [109] developed optimal trapezoidal and rectangular sections considering seepage. Explicit formulas for optimal polygonal sections considering seepage loss were developed by Chahar [110] and Swamee et al. [111-113] and for non-polygonal sections by Swamee and Kashyap [114, 115].

For curved channel sections, such as semicircular and parabolic, exact analytical solutions of seepage loss are not possible. Chahar [116] has reviewed previous research work since 1962 to address this issue. The research aimed at finding the optimal shape of a curved section that achieved minimum seepage loss. Chahar [110] developed an exact solution for seepage from a curved section whose boundary mapped along a circle onto the hodograph plane. The perimeter of this section always lied between a parabola and an ellipse with the same top width and flow depth. The author subsequently developed the properties of the optimal section that minimized flow area and seepage loss for the curved section [116].

6.8.4. Section with Smooth Top Corners

The top corners of the main channel of a compound section that is connected to the flood plains are generally smooth, but they have been assumed sharp to simplify the design. A new channel section with third-degree polynomial sides that allows the top corner of the section sides to be smooth has been proposed by Easa and Han [117]. The geometry of the section is shown in Fig. 6.7. The section side is mathematically represented by

$$y = bx + cx^2 + dx^3, \qquad (6.46)$$

where b, c, and d are the parameters, and x and y are the Cartesian coordinates. For the special case of $B_f \leq - c/3d$, where B_f is the side width at the ground level, the proposed section has sharp top corners similarly to conventional sections. The characteristics of the new section, including flow area and wetted perimeter, were developed.

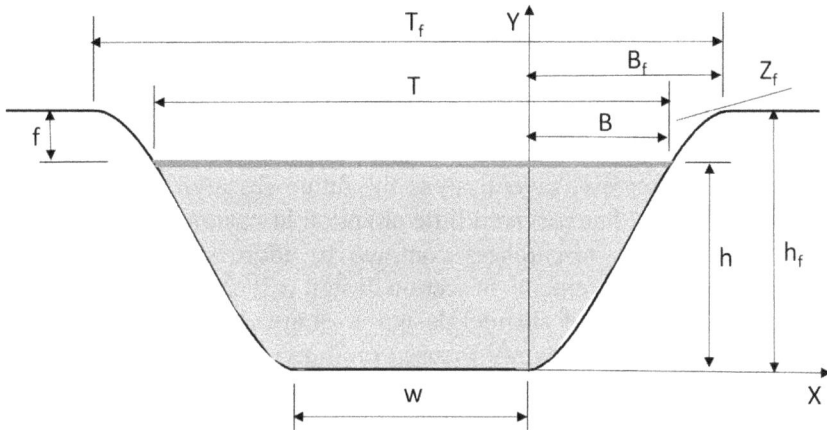

Fig. 6.7. General polynomial section with smooth top corners [117]

The special cases of the proposed polynomial section are trapezoidal, rectangular, triangular, and parabolic sections. The results showed that the proposed section was more economical than the popular parabolic section for the case of restricted side slopes. The idea of smooth top corners of a channel section is new and should be of interest to researchers and practitioners.

6.9. Concluding Remarks

This chapter has highlighted recent developments in open channel section design based on a review of 117 publications. Since 2003, 13 new section shapes (both linear and linear-curved) have been developed. Compared with conventional sections, the new sections provided less construction cost, allowed more design constraints to be accommodated, or provided the designer with more design choices. This chapter has also reviewed various design methods for rigid and flexible channels and discussed several important design considerations.

The review of the historical development of channel sections showed that conventional sections with different shapes have been developed and implemented in practice for some time. During 2003-2009, several inspiring initiatives emerged that produced new section shapes with improved performance. This has subsequently led to a surge of research that has resulted in more ideas for section shapes. This trend is strikingly similar to the historical development of the hydrologic Muskingum models presented in Chapter 7 of this book. In addition, some probabilistic methods of channel section design have been sporadically developed during the past few decades.

Based on this review, several areas for future research are suggested. First, an area that has received little attention in channel section design is uncertainty. As researchers continue to address data and model uncertainties, improvements in section design will occur. Second, multi-criteria optimization of channel design is an area that can be explored. Previous research has mainly focused on individual hydraulic and cost criteria. There is a need to quantify other criteria such as environmental and aesthetics and to consider multi-criteria in design. Recent optimization advances in this area, such as Pareto analysis, can greatly aid open channel design. Third, current open channel design normally assumes that channel longitudinal slope is given. This slope is normally determined at the planning stage based on consideration of topography and flow velocity. Future research to develop an integrated planning-design process is needed. In this process, the longitudinal slope will be a decision variable and its effect on channel performance would be directly incorporated in design. Finally, with the availability of so many open channel section shapes, future comprehensive studies are needed to evaluate thoroughly the properties of various sections and perhaps establish general guidelines regarding their implementation in practice.

This chapter has presented the current state-of-the-art in open channel design with a focus on steady uniform flow. It is hoped that the information provided will increase awareness of past and recent developments in open channel design and help researchers to further enrich our knowledge in this field.

Acknowledgements

It has been an enjoyable journey while writing this chapter. I want to thank all the researchers who have contributed to the recent developments of open channel sections. Without their fine contributions this review would not have been possible. I am especially grateful to my research collaborator, Dr. Yancheng Han of University of Jinan, China for many fruitful discussions and debates. Finally, I wish to express my special thanks to Prof. Sergey Yurish for his great support throughout the development of this chapter and the next chapter.

References

[1]. R. T. Kilgore, G. K. Cotton, Design of Roadside Channels with Flexible Linings, Hydraulic Engineering Circular No. 15, *U.S. Federal Highway Administration*, 2005.

[2]. Hydraulic Design Manual, *Texas Department of Transportation*, Austin, TX, 2016.

[3]. A. O. Akan, Open Channel Hydraulics, *Springer*, New York, 2011.

[4]. M. H. Chaudhry, Open Channel Flow, *Springer*, New York, 2008.

[5]. R. H. French, Open Channel Hydraulics, *McGraw-Hill*, New York, N.Y., 2007.

[6]. V. T. Chow, Open Channel Hydraulics, *McGraw-Hill*, New York, N.Y., 1959.

[7]. P. K. Swamee, K. G. Bhatia, Economic open channel section, *Irrig. Power*, Vol. 29, Issue 2, 1972, pp. 169-176.

[8]. C. Guo, W. Hughes, Optimal channel cross section with freeboard, *J. Irrig. Drain. Eng.*, Vol. 110, Issue 3, 1984, pp. 304-314.

[9]. A. Das, Optimal channel cross section with composite roughness, *J. Irrig. Drain. Eng.*, Vol. 126, Issue 1, 2000, pp. 68-72.

[10]. A. Jain, R. K. Bhattacharjya, S. Sanaga, Optimal design of composite channels using genetic algorithm, *J. Irrig. Drain. Eng.*, Vol. 130, Issue 4, 2004, pp. 286-295.

[11]. B. Aksoy, A. Altan-Sakarya, Optimal lined channel design, *Can. J. Civ. Eng.*, Vol. 33, Issue 5, 2006, pp. 535-545.

[12]. Y-C. Han, S. M. Easa, X-P. Gao, General explicit solutions of most economic sections and applications to trapezoidal and parabolic channels, *J. of Hydrodynamics*, 2018 (in press).

[13]. D. C. Froehlich, Width and depth-constrained best trapezoidal section, *J. Irrig. Drain. Eng.*, Vol. 120, Issue 4, 1994, pp. 828-835.

[14]. C. J. Posey, Computation of discharge including over-bank flow, *Civ. Eng.*, Vol. 37, Issue 4, 1967, pp. 62-63.

[15]. K. Shiono, D. W., Knight, Two-dimensional analytical solution for a compound channel, *Proc., 3rd Inter. Symposium on Refined Flow Modeling and Turbulence Measurements*, Japan, 1988, pp. 503-510.

[16]. J. B. Wark, P. G. Samuels, D. A. Ervine, A practical method of estimating velocity and discharge in compound channels, *Proc., Inter. Conference on River Flood Hydraulics*, London, 1990, pp. 163-172.

[17]. P. Ackers, Hydraulic design of two-stage channels, *J. Water Marit. Eng.*, Vol. 96, 1992, pp. 247-257.

[18]. P. R. Wormleaton, D. J. Merrett, An improved method of calculation for steady uniform flow in prismatic main channel/flood plain sections, *J. Hydraul. Res.*, Vol. 28, 1990, pp. 157-174.

[19]. M. F. Lambert, R. C. Myers, Estimating the discharge capacity in straight compound channels, *Water Marit. Energy*, Vol. 130, 1998, pp. 84-94.

[20]. D. Bousmar, Y. Zech, Momentum transfer for practical flow computation in compound channels, *J. Hydraul. Eng.*, Vol. 125, Issue 7, 1999, pp. 670-696.

[21]. D. A. Ervine, K. Babaeyan-Koopaei, R. H. Sellin, Two-dimensional solution for straight and meandering overbank flows, *J. Hydraul. Eng.*, Vol. 126, Issue 9, 2000, pp. 653-669.

[22]. S. M. Hosseini, Equations for discharge calculation in compound channels having homogenous roughness, *Iran. J. Sci. Technol. Trans. B*, Vol. 28, Issue 5, 2004, pp. 537-546.

[23]. A. B. MacLeod, Development of methods to predict the discharge capacity in model and prototype meandering compound channels, PhD Thesis, *University of Glasgow*, 1997.

[24]. W. Liu, C. S. James, Estimating of discharge capacity in meandering compound channels using artificial neural networks, *Can. J. Civ. Eng.*, Vol. 27, Issue 2, 2000, pp. 297-308.

[25]. A. Zahiri, A. A. Dehghani, Flow discharge determination in straight compound channels using ANN, *World Acad. Sci. Eng. Technol.*, Vol. 58, 2009, pp. 1-8.

[26]. B. Unal, M. Mamak, G. Seckin, M. Cobaner, Comparison of an ANN approach with 1-D and 2-D methods for estimating discharge capacity of straight compound channels, *Adv. Eng. Software*, Vol. 41, 2010, pp. 120-129.

[27]. H. Azamathulla, A. R. Zahiri, Flow discharge prediction in compound channels using linear genetic programming, *J. Hydrol.*, Vol. 454-455C, 2012, pp. 203-207.

[28]. M. A. Haidera, E. M. Valentine, A practical method for predicting the total discharge in mobile and rigid boundary compound channels, *Proc., Inter. Conference on Fluvial Hydraulics*, Belgium, 2002, pp. 153-160.

[29]. F. Huthoff, P. C. Roose, D. C., Augustijn, S. J. Hulscher, Interacting divided channel method for compound channel flow, *J. Hydraul. Eng.*, Vol. 134, Issue 8, 2008, pp. 1158-1165.

[30]. A. P. Mironenko, L. S. Willardson, S. A. Jenab, Parabolic canal design and analysis, *J. Irrig. Drain. Eng.*, Vol. 110, Issue 2. 1984, pp. 241-246.

[31]. G. V. Loganathan, Optimal design of parabolic canals, *J. Irrig. Drain. Eng.*, Vol. 117, Issue 5, 1991, pp. 716-735.

[32]. B. R. Chahar, Optimal design of parabolic canal section, *J. Irrig. Drain. Eng.*, Vol. 131, Issue 6, 2005, pp. 546-554.

[33]. B. R. Chahar, N. Ahmed, R. Godara, Optimal parabolic section with freeboard, *J. Indian Water Works Assoc.*, January-March 2007.

[34]. A. Laycock, Pehur high-level canal, NWFP, Pakistan, *ICE – Water Manage.*, Vol. 158, Issue 3, 2005, pp. 93-102.

[35]. A. S. Hussein, Simplified design of hydraulically efficient power-law channels with freeboard, *J. Irrig. Drain. Eng.*, Vol. 134, Issue 3, 2008, pp. 380-386.

[36]. T. S. Strelkoff, A. J. Clemmens, Approximating wetted perimeter in power-law cross section, *J. Irrig. Drain. Eng.*, Vol. 126, Issue 2, 2000, pp. 98-109.

[37]. A. A. Anwar, T. T. de Vries, Hydraulically efficient power law channels, *J. Irrig. Drain. Eng.*, Vol. 129, Issue 1, 2003, pp. 18-26.

[38]. A. R. Kacimov, Discussion of "Hydraulically efficient power law channels, by A. A. Anwar and T. T. de Vries," *J. Irrig. Drain. Eng.*, Vol. 130, Issue 5, 2004, pp. 445-446.

[39]. A. A. Anwar, D. Clarke, Design of hydraulically efficient power-law channels with freeboard, *J. Irrig. Drain. Eng.*, Vol. 131, Issue 6, 2005, pp. 560-563.

[40]. H. Carson, H. Kingman, T. Haynes, H. N. Collison, Cross-sections of sewers and diagrams showing hydraulic elements of four general types, *Engineering News,* Vol. 30, Issue 5, 1894, pp. 121-123.

[41]. G. P. Merkley, Standard horseshoe cross section geometry, *Agricultural Water Manage.*, Vol. 71, Issue 1, 2005, pp. 61-70.

[42]. W. H. Hager, Wastewater Hydraulics: Theory and Practice, *Springer*, London, U.K., 2010.

[43]. K. Babaeyan-Koopaei, E. M. Valentine, D. C. Swailes, Optimal design of parabolic-bottomed triangle canals, *J. Irrig. Drain. Eng.*, Vol. 126, Issue 6, 2000, pp. 408-411.

[44]. B. R. Chahar, S. Basu, Optimal design of curved bed trapezoidal canal sections, *ICE – Water Manage.*, Vol. 162, Issue 3, 2009, pp. 233-240.

[45]. S. O. Kurbanov, N. V. Khanov, A hydraulic analysis of the most efficient section of a power engineering channel with a polygonal profile, *Power Tech. and Eng.*, Vol. 37, Issue 4, 2003, pp. 213-216.

[46]. A. A. Uginchus, Channels: Hydraulic Design and Engineering Economy Study, *Stroiizdat*, Moscow, 1965 (in Russian).

[47]. S. O. Kurbanov, N. V. Khanov, Computation of critical depth of channels with polygonal profile, *Power Tech. and Eng.*, Vol. 38, Issue 2, 2004, pp. 42-44.

[48]. A. Abdulrahman, Best hydraulic section of a composite channel, *J. Hydraul. Eng.*, Vol. 133, Issue 6, 2007, pp. 695-697.

[49]. A. Das, Optimal design of channel having horizontal bottom and parabolic sides, *J. Irrig. Drain Eng.*, Vol. 133, Issue 2, 2007, pp. 192-197.

[50]. R. E. Horton, Separate roughness coefficients for channel bottom and sides, *Eng. News Record,* Vol. 111, Issue 2, 1933, pp. 652-653.

[51]. H. Azamathulla, R. D. Jarrett, Use of gene-expression programming to estimate Manning's roughness coefficient for high gradient streams, *Water Resour. Manage.*, Vol. 27, Issue 3, 2013, pp. 715-729.

[52]. Design of roadside channels with flexible linings, Hydraulic Engineering Circular Number 15, 3rd Ed., *Federal Highway Administration*, Washington, D.C., 2007.

[53]. D. C. Froehlich, Most hydraulically efficient standard lined canal sections, *J. Irrig. Drain Eng.*, Vol. 134, Issue 4, 2008, pp. 462-470.

[54]. S. M. Easa, Improved channel cross section with two-segment parabolic sides and horizontal bottom, *J. Irrig. Drain. Eng.*, Vol. 135, Issue 6, 2009, pp. 357-365.

[55]. A. R. Vatankhah, Discussion of "Improved channel cross section with two-segment parabolic sides and horizontal bottom, by S. M. Easa, 2009, Vol. 135, No. 3, pp. 357-365.", *J. Irrig. Drain Eng.*, Vol. 136, Issue 9, 2010, pp. 662-665.

[56]. S. M. Easa, Closure to "Improved channel cross section with two-segment parabolic sides and horizontal bottom, by S. M. Easa, 2009, Vol. 135, No. 3, pp. 357-365.", *J. Irrig. Drain Eng.*, Vol. 136, Issue 9, 2010, pp. 667-670.

[57]. A. R. Vatankhah, Semi-regular polygon as the best hydraulic section in practice (generalized solutions), *Flow Meas. Instrum.*, Vol. 38, 2014, pp. 67-71.

[58]. S. M. Easa, New and improved channel cross section with piece-wise linear or smooth sides, *Can. J. Civ. Eng.*, Vol. 38, Issue 6, 2011, pp. 690-697.

[59]. Y-C. Han, Z. Xu, S. M. Easa, S. Wang, L. Fu, Optimal hydraulic section of ice-covered open trapezoidal channel. *J. Cold Regions Eng.*, Vol. 31, Issue 3, 2017.

[60]. S. M. Easa, A. R. Vatankhah, New open channel with elliptic sides and horizontal bottom, *KSCE J. Civ. Eng.*, Vol. 18, Issue 4, 2014, pp. 1197-1204.

[61]. S. M. Easa, Versatile general elliptic open channel cross section, *KSCE J. Civ. Eng.*, Vol. 20, Issue 4, 2016, pp. 1572-1581.

[62]. Y-C. Han, S. M. Easa, Exact solution of optimum hydraulic power-law section with general exponent parameter, *J. Irrig. Drain. Eng.*, 2018 (in press).

[63]. Y-C. Han, S. M. Easa, Superior cubic channel section and analytical solution of best hydraulic properties. *Flow Meas. Instrum.*, Vol. 50, 2016, pp. 169-177.

[64]. Y-C. Han, S. M. Easa, New and improved three and one-third parabolic channel and best hydraulic section. *Can. J. Civ. Eng.*, 44, 5, 2017, pp. 387-391.

[65]. P. Monadjemi, General formulation of best hydraulic channel section, *J. Irrig. Drain. Eng.*, Vol. 120, Issue 1, 1994, pp. 27-35.

[66]. Y. K. Tung, L. W. Mays, Risk analysis for hydraulic design, *J. Hydraul. Div.*, Vol. 106, Issue 5, 1980, pp. 893-913.

[67]. M. A. Cesare, First order analysis of open channel flow, *J. Hydraul. Div.*, Vol. 117, Issue 2, 1991, pp. 242-247.

[68]. S. M. Easa, Probabilistic design of open drainage channels, *J. Irrig. Drain. Eng.*, Vol. 118, Issue 6, 1992, pp. 868-881.

[69]. S. M. Easa, Reliability analysis of open drainage channels under multiple failure modes, *J. Irrig. Drain. Eng.*, Vol. 120, Issue 6, 1994, pp. 1007-1024.

[70]. S. M. Easa, Closure to 'Probabilistic Design of Open Drainage Channels, by S. M. Easa, Vol. 118, Issue 6, 1992, 868-881', *J. Irrig. Drain. Eng.*, Vol. 120, Issue 1, 1994, pp. 228-229.

[71]. A. Das, Flooding probability constrained optimal design of trapezoidal channels, *J. Irrig. Drain. Eng.*, Vol. 133, Issue 1, 2007, pp. 53-60.

[72]. A. Das, Chance constrained optimal design of trapezoidal canals, *J. Water Resour. Plan. Manage.*, Vol. 134, Issue 3, 2008, pp. 310-313.

[73]. A. Das, Cost and flooding probability minimization based design of HBPS channel, *Water Resour. Manage.*, Vol. 24, 2010, pp. 193-238.

[74]. S. Adarsh, A.S. Sahana, Optimal design of drainage channels using probabilistic search, *ICE – Water Manage.*, Vol. 166, 2013, pp. 285-300.

[75]. M. J. Reddy, S. Adarsh, Chance constrained optimal design of composite channels using meta-heuristic techniques, *Water Resour. Manage.*, Vol. 24, 2010, pp. 2221-2235.

[76]. S. Adarsh, Modeling parametric uncertainty in optimal open channel design using FORM-PGSL coupled approach, *Stoch. Environ. Res. Risk Assess.*, Vol. 26, 2012, pp. 709-720.

[77]. S. M. Easa, Incorporating effect of roughness uncertainty into design of ice-covered channels, *J. Cold Regions Eng.*, Vol. 31, Issue 2, 2017.

[78]. A. J. Raudkivi, Loose Boundary Hydraulics, *Pergamon Press*, Great Britain, 1990.

[79]. S. M. Easa, G. Wu, A. O. Abd El Halim, M. Yu, Non-iterative method for optimal design of flexible roadside channels with bends, *ICE – Water Manage.*, Vol. 168, Issue 6, 2015, pp. 245-255.

[80]. S. K. Gupta, U. Mishra, V. P. Singh, Design of minimum cost earthen channels having side slopes riveted with different types of riprap stones and unlined bed by using particle swarm optimization, *Irrig. Drain.*, Vol. 65, 2016, pp. 319-333.

[81]. R. K. Bhattacharjya, M. G. Satish, Optimal design of a stable trapezoidal cross section using hybrid optimization techniques, *J. Irrig. Drain Eng.*, Vol. 133, Issue 4, 2007, pp. 323-329.

[82]. D. C. Froehlich, Mass angle of repose of open-graded rock riprap, *J. Irrig. Drain Eng.*, Vol. 137, Issue 7, 2011, pp. 454-461.

[83]. S. M. Easa, A. R. Vatankhah, A. O. Abd El Halim, Simplified direct method for finding optimal stable trapezoidal channel cross sections, *Inter. J. River Basin Manage.*, Vol. 9, Issue 2, 2011, pp. 85-92.

[84]. Grass-Lined Channels, Illinois Urban Manual, Practice Standard, *IEPA (Illinois Environmental Protection Agency),* Springfield, Illinois, 2010.

[85]. M. Escarameia, Y. Gasowski, R. May, Grassed drainage channels – hydraulic resistance characteristic, *Water and Maritime Eng.*, Vol. 154, Issue 4, 2002, pp. 333-341.

[86]. A. O. Akan, W. H. Hager, Design aid for grass-lined channels, *J. Hydraul. Eng.*, Vol. 127, Issue 3, 2001, pp. 236-237.

[87]. S. M. Easa, A. R. Vatankhah, Direct solutions for design of grass-lined channels, *ICE – Water Manage.*, Vol. 165, Issue 3, 2012, pp. 153-159.

[88]. A. R. Vatankhah, S. M. Easa Simplified accurate solution for design of erodible trapezoidal channels, *J. Hydro. Eng.*, Vol. 16, Issue 11, 2011, pp. 960-965.

[89]. P. K. Swamee, Normal depth equations for irrigation canals, *J. Irrig. Drain. Eng.*, Vol. 120, Issue 5, 1994, pp. 942-948.

[90]. P. K. Swamee, P. N. Rathie, Exact solutions for normal depth problem, *J. Hydraul. Res.*, Vol. 42, Issue 5, 2004, pp. 541-547.

[91]. P. K. Swamee, P. N. Rathie, Exact equations for critical depth in a trapezoidal canal, *J. Irrig. Drain Eng.*, Vol. 131, Issue 5, 2005, pp. 474-476.

[92]. Y. F. Zhao, Q. Lu, K. D. Zhang, An approximate formula for calculating water depth of uniform flow in circular cross section, *J. Northwest Agriculture and Forestry*, Vol. 36, Issue 5, 2008, pp. 225-228.

[93]. B. Achour, M. Khattaoui, Computation of normal and critical depths in parabolic cross sections, *Open Civ. Eng. J.,* Vol. 2, Issue 1, 2008, pp. 9-14.

[94]. R. V. Raikar, M. S. Shiva Reddy, G. K. Vishwanadh, Normal and critical depth computations for egg-shaped conduit sections, *Flow Meas. Instrum.,* Vol. 21, Issue 3, 2010, pp. 367-372.

[95]. A. R. Vatankhah, S. M. Easa, Explicit solutions for critical and normal depths in channels with different shapes, *Flow Meas. Instrum.*, Vol. 22, Issue 1, 2011, pp. 43-49.

[96]. J. L. Liu, Z. Z. Wang, C. J. Leng, Y. F. Zhao, Explicit equations for critical depth in open channels with complex compound cross sections, *Flow Meas. Instrum.*, Vol. 24, Issue 1, 2012, pp. 13-18.

[97]. A. R. Vatankhah, Direct solutions for normal and critical depths in standard city-gate sections, *Flow Meas. Instrum.*, Vol. 28, 2012, pp. 16-21.

[98]. A. R. Vatankhah, Explicit solutions for critical and normal depth sin trapezoidal and parabolic open channels, *Ain Shams Eng. J.*, Vol. 4, Issue 1, 2013, pp. 17-23.

[99]. A. R. Vatankhah, Critical and normal depths in semi-elliptical channels, *J. Irrig. Drain Eng.*, Vol. 141, Issue 10, 2015.

[100]. A. R. Vatankhah, Normal depth in power-law channels, *J. Hydrol. Eng.*, Vol. 20, Issue 7, 2015.

[101]. P. K. Swamee, P. N. Rathie, Normal depth equations for wide rectangular and triangular open-channel sections involving Lambert's W function, *J. Hydraul. Eng.*, Vol. 18, Issue 3, 2012, pp. 252-257.

[102]. P. K. Swamee, P. N. Rathie, Normal depth equations for parabolic open channel sections, *J. Irrig. Drain. Eng.*, 2016, Vol. 142, Issue 6, pp. 1-3.

[103]. Y. Li, Z. Gao, Explicit solution for normal depth of parabolic section of open channels, *Flow Meas. Instrum.*, Vol. 38, 2014, pp. 36-39.

[104]. J. Liu, Z. Wang, Equations for critical and normal depths of city-gate sections, *ICE – Water Manage.*, Vol. 166, Issue 4, 2013, pp. 199-206.

[105].   M. A. Perez, X. Fang, C. G. Butler, Compute critical and normal depths of arch and elliptical pipes, *J. Irrig. Drain Eng.*, Vol. 141, Issue 9, 2015.

[106].   M. Elhakeem, Explicit solution for flow depth in open channels of trapezoidal cross-sectional, *J. Irrig. Drain. Eng.*, Vol. 143, Issue 7, 2017, pp. 1-11.

[107].   Design and Construction Manual: Canals and Related Structures, Design Supplement No. 3, *U.S. Bureau of Reclamation*, 1952.

[108].   V. T. Chow, Open Channel Hydraulics, *McGraw-Hill*, New York, 1973.

[109].   A. R. Kacimov, Seepage optimization for trapezoidal channel, *J. Irrig. Drain. Eng.*, Vol. 118, Issue 4, 1992, pp. 520-526.

[110].   B. R. Chahar, Analytical solution to seepage problem from a soil channel with a curvilinear bottom, *Water Resour. Res.*, Vol. 42, Issue 1, 2006, W01403.

[111].   P. K. Swamee, G. C. Mishra, B. R. Chahar, Comprehensive design of minimum cost irrigation canal sections, *J. Irrig. Drain. Eng.*, Vol. 126, Issue 5, 2000, pp. 322-327.

[112].   P. K. Swamee, G. C. Mishra, B. R. Chahar, Design of minimum seepage loss canal sections, *J. Irrig. Drain. Eng.*, Vol. 126, Issue 1, 2000, pp. 28-32.

[113].   P. K. Swamee, G. C. Mishra, B. R. Chahar, Design of minimum water-loss canal sections, *J. Hydraul. Res.*, Vol. 40, Issue 2, 2002, pp. 215-220.

[114].   P. K. Swamee, D. Kashyap, Design of minimum seepage loss non-polygon canal sections, *J. Irrig. Drain. Eng.*, Vol. 127, Issue 2, 2001, pp. 113-117.

[115].   P. K. Swamee, D. Kashyap, Design of minimum seepage loss non-polygonal canal sections with drainage layer at shallow depth, *J. Irrig. Drain. Eng.*, Vol. 130, Issue 2, 2004, pp. 166-170.

[116].   B. R. Chahar, Optimal design of a special class of curvilinear bottom channel section, *J. Hydraul. Eng.*, Vol. 133, Issue 5, 2007, pp. 571-576.

[117].   S. M. Easa, Y-C. Han, New polynomial open channel cross section with smooth top corners, Research Report, Department of Civil Engineering, *Ryerson University*, CE-RR-01, Ontario, Canada, 2018.

Chapter 7

Channel Flood Routing: Review of Recent Hydrologic Muskingum Models

Said M. Easa

7.1. Introduction

Flood routing models are used to estimate the outflow hydrograph given the inflow hydrograph. The models are classified as hydraulic and hydrologic models [1]. Hydraulic models require measurements of both flow depth and discharge using elaborate stream gauging. These models, which are based on the continuity and motion equations, are more complicated and difficult to use, but they reflect flood-routing conditions more accurately. Hydrologic models, on the other hand, require only discharge measurements since model parameters are assumed to capture the overall flood-propagating characteristics. As such, these models are particularly useful at the initial planning stage where the gauging system is still underdeveloped or insufficient for precise measurements. One popular hydrologic flood routing model is Muskingum model which is based on the continuity and storage equations. The model, first postulated in 1938 by McCarthy [2] of the U.S. Corps of Engineers, has been frequently used for flood routing in natural channels because of its simplicity. Hydrologic flood routing can be performed for reservoirs and channels. This chapter focuses on hydrologic Muskingum models for channel routing. In total, 69 publications were reviewed and presented with a historical perspective.

Significant research has been conducted on channel routing during the past five years that has resulted in the development of many new models with different structures and better accuracy. The models are based on an alternative thinking by Easa [3] that has focused on modifying

Said M. Easa
Department of Civil Engineering, Ryerson University, Toronto, Ontario, Canada

parameter and *model structures*, unlike the traditional approach that had focused on improving *solution algorithms*. Motivation for this alternative thinking and a review of 34 new models with modified parameter/model structures are presented in this chapter.

The application of hydrologic Muskingum flood models involves two phases: *calibration* phase and *forecasting* phase (Fig. 7.1). In the calibration phase, data related to observed inflow and outflow hydrographs are used to estimate model parameters. In the forecasting phase, the calibrated model is used to forecast the outflow hydrograph of a flood given its inflow hydrograph. There are implications of the calibrated model on the forecasting phase that are related to accuracy and legal aspects. In addition, preliminary guidelines regarding selection of model type and routing procedures have been developed. These and other practical considerations are highlighted in this chapter.

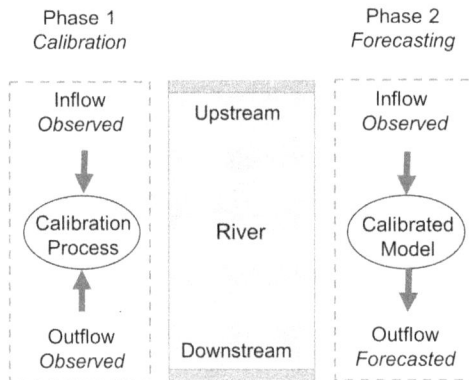

Fig. 7.1. Phases of hydrologic Muskingum model application: calibration and forecasting.

This chapter is organized as follows. Section 7.2 presents a historical perspective of the development of hydrologic Muskingum models, including the motivation for the surge of recent research. Section 7.3 reviews the original Muskingum models (1959-2013). Section 7.4 reviews two new structurally-modified Muskingum models (2013-2014) that have started an alternative thinking for improving model performance. Section 7.5 reviews subsequently developed models (2014-present). Section 7.6 describes common routing procedures of model calibration. Section 7.7 discusses practical considerations and Section 7.8 presents concluding remarks.

7.2. Historical Perspective

Hydrologic flood routing models have been developed more than 50 years ago. The work started with a three-parameter (3P) nonlinear Muskingum model with flow exponent parameter α in 1959 by Chow [4], followed by a 3P model with weighted-flow exponent parameter β in 1978 by Gill [5]. Gill's model has become more popular as it is easier to calibrate and has shown to produce better model performance. During the period 1978-2013, researchers have focused on improving the performance of Gill's model by modifying its *solution algorithm*. However, such efforts have produced little success in recent years as illustrated in Easa [3]. For example, in 1978 the sum of the squared deviations between predicted and observed outflows (*SSQ*) was 145.70 by Gill [5]. Then, in 1985 *SSQ* was substantially reduced to 45.54 by Tung [6]. Subsequently, in 1997 *SSQ* was slightly reduced to 38.24 by Mohan [7]. Since 1997, nine solution algorithms have been developed by Mohan [7], Kim et al. [8], Geem [9], Chu and Chang [10], Luo and Xie [11], Barati [12], Geem [13], Xu et al. [14], Orouji et al. [15], and Karahan et al. [16]. However, the algorithms produced only an improvement in model performance of less than 0.4 %, where *SSQ* was reduced from 36.89 to 36.76, as shown in Table 7.1. In addition, during the period 2011-2013 four new algorithms practically produced no improvements.

Table 7.1. Performance of Gill's 3P model using 12 solution algorithms using Wilson's data (1978-2013).

Algorithm	Reference	Year	*SSQ*
Segmented least-squares	Gill [5]	1978	145.70
HJ+DFP [a]	Tung [6]	1985	45.54
Genetic	Mohan [7]	1997	38.24
Harmony search (HS)	Kim et al. [8]	2001	36.78
BFGS [b]	Geem [9]	2006	36.77
Particle swarm optimization	Chu and Chang [10]	2009	36.89
Immune clonal selection	Luo and Xie [11]	2010	36.80
Nelder-Mead simplex	Barati [12]	2011	36.76
Parameter-setting-free HS	Geem [13]	2011	36.77
Differential evolution	Xu et al. [14]	2012	36.77
Meta Heuristic	Orouji et al. [15]	2013	36.77
HS-BFGS (hybrid)	Karahan et al. [16]	2013	36.77

[a] Hooke-Jeeves pattern search and the Davidon-Fletcher-Powell algorithm.
[b] Broyden-Fletcher-Goldfarb-Shanno.

For this reason, alternative thinking to improve model performance by modifying *parameter* and *model structures*, instead of solution algorithms, has been proposed by Easa [17, 18] who developed two models published in 2013 and 2014 that substantially improved model performance. In the first model, a variable exponent parameter was used and in the second model the structure of the 3P Muskingum model was modified resulting in a four-parameter model. The first model reduced *SSQ* to from 36.77 to 27.39 using two inflow sub-regions [17] and the second model reduced *SSQ* to 6.76 [18]. These values represent reductions in *SSQ* of 26 % and 80 %, respectively, compared with the traditional 3P model. Following the development of these two models, other structurally-modified Muskingum models were developed.

Three distinct types of hydrographs have been commonly used in the literature for the calibration of Muskingum models [19-21]: single-peak, smooth (SPS) hydrograph, single-peak, non-smooth (SPN) hydrograph, and multi-peak (MP) hydrograph, as shown in (Fig. 7.2).

7.3. Original Muskingum Models (1959-2013)

The flood routing procedure is based on the hydrologic continuity and storage equations. The continuity equation is given by

$$\frac{dS}{dt} = I - Q \ \text{(Continuity)}, \qquad (7.1)$$

where dS/dt is the rate of change of the storage volume with respect to time t and S, I, and Q are the simultaneous amounts of storage, inflow, and outflow, respectively. To simplify the notation, the time-index subscript t is omitted in model presentation. The storage is a function of I, Q, and several other parameters and can be expressed as

$$S = \text{f}(I, \ Q, \ K, \ w, \ \alpha, \ \beta) \ \text{(Storage)}, \qquad (7.2)$$

where K is the storage parameter, w is the weighting parameter that represents the inflow-outflow relative effects on the storage, and α and β are other parameters that account for the flow and storage characteristics at the upstream and downstream cross sections. All Muskingum models use the same continuity equation, but they vary in the functional form of the storage equation. The storage equation is referred to throughout as Muskingum model.

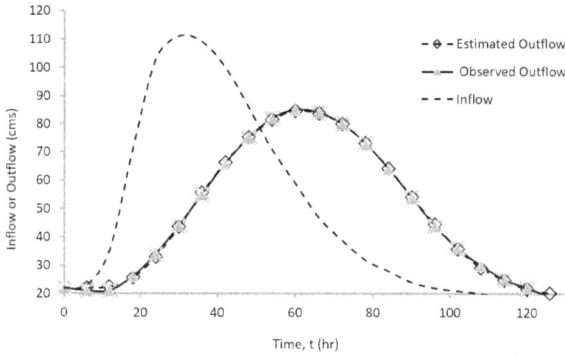

(a) Single-peak, smooth hydrograph [19].

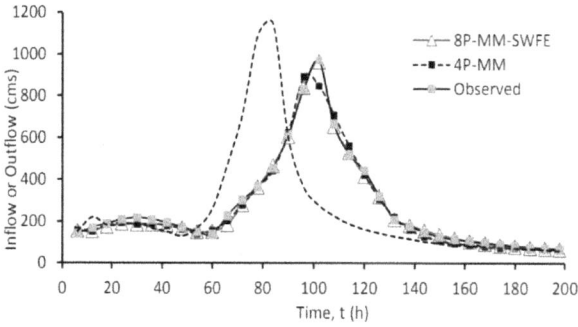

(b) Single-peak, non-smooth hydrograph [20].

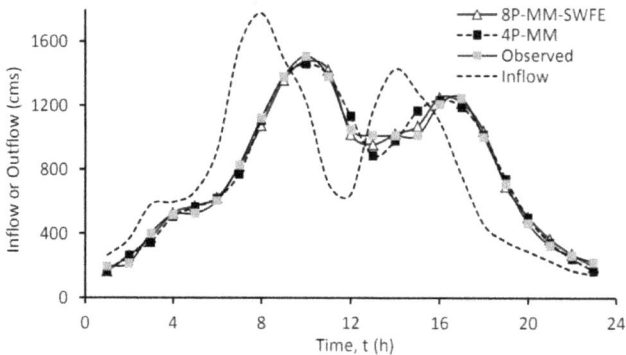

(c) Multi-peak hydrograph [21].

Fig. 7.2. Common types of hydrographs used in the calibration of hydrologic Muskingum models.

261

The earliest nonlinear Muskingum model with a flow exponent parameter was derived in 1959 by Chow [4]. This derivation is presented here as it forms the basis for subsequent developments. Chow assumed the storage to be a linear function of the weighted inflow and outflow. That is,

$$S = [w\, S_{in} + (1-w)\, S_{out}] \tag{7.3}$$

Assuming that the flow and storage characteristics at the upstream and downstream cross sections are constant, then

$$I = ay^n \tag{7.4}$$

$$Q = ay^n \tag{7.5}$$

$$S_{in} = by^m \tag{7.6}$$

$$S_{out} = by^m, \tag{7.7}$$

where y is the flow depth, a and n reflect the discharge-depth characteristics and b, m reflect the storage-depth characteristics of the reach. Eliminating y from Eqs. (7.4) and (7.6), and Eqs. (7.5) and (7.7) yields

$$S_{in} = b\left(\frac{I}{a}\right)^{m/n} \tag{7.8}$$

$$S_{out} = b\left(\frac{Q}{a}\right)^{m/n} \tag{7.9}$$

Substituting for S_{in} and S_{out} from Eqs. (7.8) and (7.9) into Eq. (7.3), then

$$S = K\left[wI^\alpha + (1-w)I^\alpha\right], \quad (3\text{P-C1}), \tag{7.10}$$

where $K = (b/a^\alpha)$ and $\alpha = m/n$. The designation 3P-C1 means the number of model parameters in model function (*NP*) is 3, all are constant, with 1 indicating model number. If there is only one model with the same *NP*, model number is omitted. If one of the parameters is variable, V is used instead of C. If the model includes a lateral flow parameter, L is added. For example, 10P-V-L means *NP* = 10, one or more parameter is variable, and one is the lateral flow parameter.

In 1985, another nonlinear model with two different parameters for inflow and outflow was proposed by Gavilan and Houck [22] as

$$S = K\left[wI^{\alpha_1} + (1-w)I^{\alpha_2}\right], \quad \text{(4P-C1)} \tag{7.11}$$

Due to the difficulty in solving nonlinear optimization models at the time, the parameter α of Eq. (7.10) has been assumed equal to 1, resulting in the following two-parameter linear model

$$S = K\left[wI + (1-w)Q\right], \quad \text{(2P-C)} \tag{7.12}$$

In 1978, to improve the performance of the linear model of Eq. (7.12), Gill [5] added a new exponent parameter β to the weighted storage of the model. The resulting three-parameter nonlinear model is given by

$$S = K\left[wI + (1-w)Q\right]^{\beta}, \quad \text{(3P-C2)} \tag{7.13}$$

This model produced better performance than the 3P-C1 model of Eq. (7.10) and has become very popular in practice. Note that the model of Eq. (7.13) had not been derived by Gill [5]. The parameter β was added in an ad-hoc manner to account for the nonlinearity of flood wave.

In actual flood events, lateral flow exists in the river reach. To model this type of flow, the continuity equation, Eq. (7.1), is modified as

$$\frac{dS}{dt} = (1 + \varphi)I - Q, \tag{7.14}$$

where φ is the lateral flow parameter. A negative value of φ means 'outflow' with water flowing from the river to the floodplains, while a positive value means 'in-flow' with water flowing from the intervening catchment rainfall and tributaries to the channel. In 1985, O'Donnell [23] was the first researcher to highlight the importance of incorporating lateral flow in flood routing. He modified the 2P-C linear model of Eq. (7.12) to include a lateral flow parameter, as follows

$$K\left[w(1 + \varphi)I + (1-w)Q\right], \quad \text{(3P-C-L)} \tag{7.15}$$

In this model, it is assumed that the lateral flow is linearly proportional to the inflow and linearly distributed along the reach. The basic model of Eq. (7.15) was subsequently extended by several researchers, including O'Donnell [24], Khan [25], and Choudhury et al. [26]. More recent developments are addressed in Section 7.5.4.

7.4. New Inspiring Initiatives (2013-2014)

As previously mentioned, during the period 1985-2013 researchers developed many solution algorithms to improve the performance of the 3P-C2 model, but the improvements were very small. This has motivated the writer to initiate an alternative thinking to improve model performance by modifying parameter and model structures. In this regard, two structurally-modified Muskingum models were developed and published in 2013 and 2014 and are described next.

7.4.1. Model with Variable Exponent Parameter

Based on an idea by Azania and Zahraie [27] who considered variable parameters to estimate daily discharges, in 2013 Easa [17] developed a nonlinear Muskingum model with a variable exponent parameter that discretely varied with the inflow. The model takes the following form

$$S = K\left[wI + (1-w)Q\right]^{\beta(u_t)}, \ [(2+NL)\text{P-V}], \qquad (7.16)$$

where $\beta(u_t)$ is the variable exponent parameter for time interval t, assumed to vary with the inflow level and NL is the number of inflow sub-regions (levels). An inflow-dividing parameter for sub-region i is defined as

$$r_i = \frac{u_i}{I_{max}}, \text{ and } r_{i+1} > r_i, \text{ for } i = 2, 3, \ldots, NL - 1, \qquad (7.17)$$

where u_i is the inflow for sub-region i and I_{max} is the maximum inflow. For sub-region i, the exponent parameter is β_i, where $i = 1, 2, \ldots, NL$. The number of model parameters equals $(2 + NL)$.

Let the number of time intervals used for flood routing be denoted by n. Then, the exponent parameter for time interval t, p_t, is defined as

$$p_t = \begin{cases} \beta_1 \text{ if } I_1 \leq u_1 \\ \beta_2 \text{ if } u_1 < I_1 \leq u_2 \\ \quad \cdot \\ \quad \cdot \\ \quad \cdot \\ \beta_L \text{ if } I_1 > u_{L-1} \end{cases} \qquad (t = 1, 2\ldots, n) \qquad (7.18)$$

The effect of the number of inflow sub-regions on model performance is shown in Table 7.2.

There were four discussions of Easa [17] by Vatankhah [28], Hamedi et al. [29], Luo and He [30], and Karahan [31]. The discussers have suggested several ideas for modifying the structure of the proposed variable exponent parameter. Some ideas were useful and others had adverse implications for model forecasting. The closure to these discussions has also presented highlights of the alternative thinking to improve model performance [3].

Table 7.2. Effect of the number of inflow sub-regions on model performance.

NL	SSQ	SSQ Reduction (%)[b]	Min β_t	Max β_t	K	w
1[a]	36.77	0	1.868	1.868	0.518	0.287
2	27.39	25.5	1.900	1.928	0.453	0.285
3	26.09	29.1	1.911	1.939	0.425	0.269
4	25.37	31.0	1.889	1.916	0.459	0.257
5	24.88	32.3	1.880	1.906	0.483	0.266

[a] This corresponds to the case of a constant exponent parameter β.
[b] The reduction is calculated in comparison with the results of a constant exponent parameter ($SSQ = 36.77$).

Karahan [31] and Vatankhah [32] have incorrectly labelled the new model of Eq. (7.16) as a 12P model and criticized the model for having too many parameters! The fact is that the number of parameters of the model depends on the number of inflow sub-regions NL, as clearly stated in the paper. For example, for $NL = 2$ the model has four parameters. As noted by Easa [3], the main purpose of the paper was to introduce a new concept related to the structure of the exponent parameter, and different values of $NL = 2$ to 10 were used to illustrate its effect on model performance. It was shown that considering the exponent parameter as a variable can substantially improve model performance. It was found that as the number of inflow sub-regions increases, model performance improves, but the improvement was small beyond $NL = 5$. Even for $NL = 2$ (4P-V model), SSQ was substantially reduced to 27.39, compared with the best $SSQ = 36.77$ of the traditional 3P-C2 model. This represents a reduction of 26 %. Clearly, this was an important development after 12 years of modifying the solution algorithms and improving SSQ by only 0.4 %, as shown previously in Table 7.1.

7.4.2. Model with 4 Constant Parameters

In 2014, Easa [18] modified model structure and derived a new model with four constant parameters. Given the complexity of the unsteady flow and the empirical nature of the model, the storage in a channel was expressed as a power function of the weighted storage such that

$$S = \left[w S_{in} + (1 - w) S_{out} \right]^{\beta} \qquad (7.19)$$

Substituting for S_{in} and S_{out} from Eqs. (7.8) and (7.9) into Eq. (7.19) yields a four-parameter model, as follows

$$S = K \left[w I^{\alpha} + (1 - w) Q^{\alpha} \right]^{\beta}, \quad \text{(4P-C2)}, \qquad (7.20)$$

where $K = (b/a^{\alpha})^{\beta}$. As noted, this model includes both parameters of the previous nonlinear models, 3P-C1 and 3P-C2. The added model parameter provided more flexibility in fitting the observed outflow hydrograph. For Wilson's data, the 4P-C2 model reduced *SSQ* by almost 80 % compared with Gill's 3P-C2 model.

Easa [18] has also highlighted the need for further research to explore implementation of the proposed model (or its extended versions) for more complex flood reaches with multiple tributaries connecting at a common confluence or with lateral flow.

In 2016, Moghaddam et al. [33] applied the particle-swarm optimization (PSO) algorithm to estimate the parameters of the new four-parameter model of Eq. (7.20). The results showed that the PSO algorithm improved *SSQ* for SPN and MP hydrographs, compared with the generalized reduced gradient (GRG) algorithm used by Easa [18].

Karahan [34] criticized the extra parameter of the 4P-C2 model of Eq. (7.20) by saying "*increasing the number of parameters also increases the model dependence on data and reduces the prediction power. Furthermore, it is also known that increasing the number of parameters will complicate the numerical solution.*" These comments are clearly injudicious in light of the author's 10-parameter model [31].

7.5. Following Muskingum Models (2014-Present)

Since the developments of the two structurally-modified models by Easa [17, 18], at least 34 new Muskingum models with modified model or parameter structure have been developed. The models are divided into five categories: (1) models with constant parameters, (2) models with discrete variable parameters, (3) models with continuous variable parameters, (4) models with lateral flow, and (5) multi-criteria models.

7.5.1. Models with 5-7 Constant Parameters

Following the publication of the 4P-C2 model by Easa [18], several models with more constant parameters have been developed. In a discussion of the preceding paper in 2014, Barati et al. [35] followed a similar derivation of the 4P-C2 and assumed that the upstream and downstream sections had different characteristics, where Eqs. (7.8) and (7.9) become

$$S_{in} = b_1 \left(\frac{I}{a_1} \right)^{m_1/n_1} \tag{7.21}$$

$$S_{out} = b_2 \left(\frac{I}{a_2} \right)^{m_2/n_2} \tag{7.22}$$

Substituting Eqs. (7.21) and (7.22) into Eq. (7.19) yields

$$S = \left[\gamma_1 I^{\alpha_1} + \gamma_2 Q^{\alpha_2} \right]^{\beta}, \text{(5P-C1)}, \tag{7.23}$$

where $\gamma_1 = w b_1 / \left(a_1 \right)^{m_1/n_1}$, $\alpha_1 = m_1 / n_1$, and $\gamma_1 = (1 - w) b_2 / \left(a_2 \right)^{m_2/n_2}$. The model parameters are γ_1, α_1, γ_1 α_2, and β. Barati et al. [35] have shown that their model improved *SSQ* by 29 % compared with the 4P-C2 model.

In his reply, Easa [35] refined the 5P-C1 model of Eq. (7.23) to preserve the physical significance of the original parameters, and proposed the following six-parameter model

$$S = \left[w K_1 I^{\alpha_1} + (1 - w) K_2 Q^{\alpha_2} \right]^{\beta}, \text{(6P-C1)} \tag{7.24}$$

267

In 2014, Vatankhah [36] presented a 5P-C2 model in a discussion, as follows

$$S = K\left[wI^{\alpha_1} + (1-w)Q^{\alpha_2}\right]^{\beta}, \text{(5P-C2)} \qquad (7.25)$$

This model, which is very similar to the new 4P-C2 model by Easa [18], has no theoretical basis as it cannot be derived from basic principles. In addition, there are serious ethical issues associated with this model that are discussed in Section 7.7.4.

In 2015, Haddad et al. [37] proposed a seven-parameter model as

$$S = K\left[wC_1I^{\alpha_1} + (1-w)C_2Q^{\alpha_2}\right]^{\beta}, \text{(7P-C)} \qquad (7.26)$$

The model was derived from basic principles. The authors assumed the difference in morphological changes between upstream and downstream cross sections of a river reach is captured by varying the a and n coefficients of Eqs. (7.8) and (7.9). The model was solved using a hybrid algorithm of shuffled-frog leaping algorithm and Nelder-Mead simplex.

In 2016, Niazkar and Afzali [38] proposed a six-parameter model as

$$S = K\left[x_1I^{\alpha_1} + x_2Q^{\alpha_2}\right]^{\beta}, \text{(6P-C2)} \qquad (7.27)$$

Similar to the model of Eq. (7.26) by Haddad et al. [37], this model can be derived from basic principles. The two parameters x_1 and x_2 are equivalent to the corresponding three parameters of Eq. (7.26). The model was solved using modified honey-bee mating optimization.

7.5.2. Models with Discrete Variable Parameters

In 2014, a discrete variable exponent parameter of the 4P-C2 model has been proposed by Easa [3] as

$$S = K\left[wI^{\alpha} + (1-w)Q^{\alpha}\right]^{\beta(u)}, \text{(5P-V)}, \qquad (7.28)$$

where $\beta(u)$ is a variable exponent parameter for time interval t represented by a two-step function. The model parameters are K, w, α,

β_1, and β_2. Compared with the model of Eq. (7.20), this 5P-V model introduced a fifth parameter by disaggregating the exponent parameter β, unlike other models that add a parameter by using two different flow exponent parameters.

For comparison, the three models 4P-C2 of Eq. (7.20), 5P-C1 of Eq. (7.23), and 5P-V of Eq. (7.28) were applied by Easa [35] to the three hydrographs of Fig. 7.2. The results are shown in Table 7.3. As noted, the 5P-V model reduced *SSQ* by 2 %, 28 %, and 14 % compared with the 5P-C1 model for SPS, SPN, and MP hydrographs, respectively. These results indicate that increasing the number of parameters through the exponent parameter is more effective than that through the flow parameter.

In 2015, another model in which all parameters of the 4P-C2 model of Eq. (7.20) were made variables has been proposed by Easa [40], as follows

$$S = K(u)[w(u)I^{\alpha(u)} + (1 - w(u))Q^{\alpha(u)}]^{\beta(u)}, \quad (8\text{P-V}), \qquad (7.29)$$

where $K(u)$, $w(u)$, $\alpha(u)$, and $\beta(u)$ are variable parameters represented by two-step discontinuous functions and u is the dimensionless inflow variable given by Eq. (7.17). An example of modelling a two-step function is shown in Fig. 7.3. The model of Eq. (7.29) has eight parameters and produces a wide array of 15 special models with 4 to 7 parameters, as shown in Table 7.4.

Table 7.3. Performance of the 4P and 5P Muskingum models [35].

Example	SSQ^c			Improvement[b] (%)	Parameters of 5P Model with VEP					
	Four-parameter model (Easa, 2014)	Discussers' five-parameter model	Five-parameter model with VEP		K	w	α	β_1	β_2	v_1
Example 1	7·67	5·44	5·35	2	0·971	0·264	0·356	-[a]	-	-
Example 2	32 299	32 056	23 069	28	0·942	0·338	1·283	1·133	1·154	0·289
Example 3	73 379	69 727	59 709	14	0·064	0·267	1·037	1·439	1·426	0·393

[a] For this example, a continuous function of the exponent parameter provided better results, $\beta(u_j) = a + b \log(1.1 - u_j)$, where $a = 4.830$ and $b = -0.052$. Examples 1, 2, and 3 correspond to the SPS, SPN, and MP hydrographs.
[b] Improvement of the 5P-V, compared with Barati et al.'s 5P-C model.
[c] All results include *SSQ* for $t = 0$, which equals $(I_0 - Q_0)^2$.

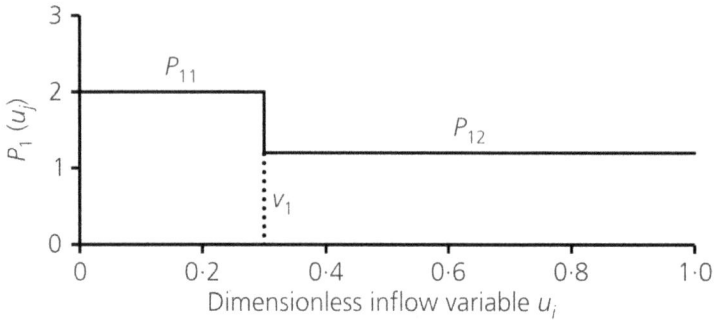

Fig. 7.3. Example of two-step function of $K(u_j)$ parameter [40].

Table 7.4. Special models of the general 8P-V model [40].

Group and Model	Storage Equation
Four-parameter group	
4P-MM	$S_j = K[wI_j^\alpha + (1 - w)Q_j^\alpha]^\beta$
Five-parameter group	
5P-MM-S	$S_j = K(u_j)[wI_j^\alpha + (1 - w)Q_j^\alpha]^\beta$
5P-MM-W[b,c]	$S_j = K[w(u_j)I_j^\alpha + [1 - w(u_j)]Q_j^\alpha]^\beta$
5P-MM-F	$S_j = K[wI_j^{\alpha(u_j)} + (1 - w)Q_j^{\alpha(u_j)}]^\beta$
5P-MM-E[a]	$S_j = K[wI_j^\alpha + (1 - w)Q_j^\alpha]^{\beta(u_j)}$
Six-parameter group	
6P-MM-SW	$S_j = K(u_j)[w(u_j)I_j^\alpha + [1 - w(u_j)]Q_j^\alpha]^\beta$
6P-MM-SF	$S_j = K(u_j)[wI_j^{\alpha(u_j)} + (1 - w)Q_j^{\alpha(u_j)}]^\beta$
6P-MM-SE	$S_j = K(u_j)[wI_j^\alpha + (1 - w)Q_j^\alpha]^{\beta(u_j)}$
6P-MM-WF	$S_j = K[w(u_j)I_j^{\alpha(u_j)} + [1 - w(u_j)]Q_j^{\alpha(u_j)}]^\beta$
6P-MM-WE[a,c]	$S_j = K[w(u_j)I_j^\alpha + [1 - w(u_j)]Q_j^\alpha]^{\beta(u_j)}$
6P-MM-FE[b]	$S_j = K[wI_j^{\alpha(u_j)} + (1 - w)Q_j^{\alpha(u_j)}]^{\beta(u_j)}$
Seven-parameter group	
7P-MM-SWF	$S_j = K(u_j)[w(u_j)I_j^{\alpha(u_j)} + [1 - w(u_j)]Q_j^{\alpha(u_j)}]^\beta$
7P-MM-SWE[a,b,c]	$S_j = K(u_j)[w(u_j)I_j^\alpha + [1 - w(u_j)]Q_j^\alpha]^{\beta(u_j)}$
7P-MM-SFE	$S_j = K(u_j)[wI_j^{\alpha(u_j)} + (1 - w)Q_j^{\alpha(u_j)}]^{\beta(u_j)}$
7P-MM-WFE	$S_j = K[w(u_j)I_j^{\alpha(u_j)} + [1 - w(u_j)]Q_j^{\alpha(u_j)}]^{\beta(u_j)}$

[a] Best model within the group for SPS hydrograph.
[b] Best model within the group for SPN hydrograph.
[c] Best model within the group for MP hydrograph.

In 2016, Afzali [41] evaluated the linear Muskingum model where the two parameters (K and w) were considered inflow-based variable parameters, similar to Eqs. (7.17) and (7.18). The routing procedure was based on the finite-difference method. The author's model reduced SSQ by up to 50 % compared with the original Muskingum model. The inflow region was divided into two and three sub-regions ($NL = 2$ and 3), where each part has its unique parameter values. The authors divided the inflow region using an additional dividing non-dimensional parameter. Following this success, in 2017, Niazkar and Afzali [42] evaluated a new nonlinear variable-parameter Muskingum model, given by

$$S = K(u)[w(u)I + (1 - w(u))Q]^{\beta(u)}, \; [(3NL)\text{P-V}], \qquad (7.30)$$

where the functions of the variable parameters are defined for the specified inflow sub-regions. The performance of 14 variable- parameter models was evaluated in a single routing application. The results showed that the proposed method significantly improved calibration statistics compared with the best three-constant parameter model. The authors concluded that among the one-variable parameter models, the best results are obtained when w is considered as a variable parameter. In addition, among the two-variable parameter models, much better results were obtained when w and K were considered as variable parameters. The finding regarding the importance of making w a variable parameter is consistent with the results of Easa [40].

7.5.3. Models with Continuous Variable Parameters

In 2014, Vatankhah [28] proposed a continuous exponent parameter in a discussion of Easa [17]. The discusser assumed each time interval j had a specific exponent parameter, where $j = 1, 2, \ldots, N$ (number of time intervals). Then, he linked these interval-specific parameters using a function, $m(t)$, that involved five coefficients, as follows

$$m(t) = a_1 + a_2 sin\left[a_3 + a_4 (t + 1)^{a_5} \right], \qquad (7.31)$$

where the angle of the *sine* function is considered in radian, and a_1 to a_5 are constant coefficients that are determined using optimization. In the closure, Easa [3] showed that such a continuous exponent parameter is just an academic exercise as its validity in the forecasting phase is questionable. This continuous exponent parameter did not capture any characteristic of the flood being simulated as it was merely a function of time and did not carry the flood signature. To address this issue, Easa [3] presented a continuous exponent function of the dimensionless inflow-based variable as follows

$$m\left(u_j\right) = a + be^{-e^{cu_j}}, \tag{7.32}$$

where $m(u_j)$ is the exponent parameter for time interval j, a, b, and c are coefficients to be determined using optimization, and u_j is the dimensionless inflow variable (0 to 1), which is given by

$$u_j = \frac{I_j}{I_{max}}, \tag{7.33}$$

where I_j is the inflow for time interval j and I_{max} is the maximum inflow during the routing period. Note that Eq. (7.32) could be increasing or decreasing function depending on the parameter values. Thus, the calibrated function can reveal a trend between the exponent parameter and the inflow, and carry this trend to the forecasting phase.

In a subsequent paper in 2017, Vatankhah [32] acknowledged the limitation of the earlier continuous function of Eq. (7.31) and proposed a continuous inflow-based exponent parameter that further improved model performance, as follows

$$m\left(z_t\right) = m_0 + a\,\sin\left(b + z_t^c\right), \tag{7.34}$$

where $m(z_t)$ is the continuous exponent parameter at time t, where the angle of the sine function is in radians, m_0, a, b, and c are parameters to be determined using optimization, $z_t = I_t / I_e$ is the dimensionless inflow variable, and I_e is the effective inflow parameter, which is determined using optimization.

7.5.4. Models with Lateral Flow

In 2014, in a discussion of the model with variable exponent parameter [17], Karahan [31] presented a model with lateral flow in which the flood period is divided into two sub-periods with respect to the peak of the inflow hydrograph, such that

$$S = K(u)[w(u)(1 + \varphi(u))\,I + (1 - w(u))Q]^{\beta(u)}, \text{ (8P-V-L)} \tag{7.35}$$

The parameters of the two sub-periods are $(K_1, w_1, \varphi_1, \beta_1)$ and $(K_2, w_2, \varphi_2, \beta_2)$. The model has two more time weighting parameters (θ_1, θ_2), bringing the total number of parameters to 10. This approach requires manually analyzing the rising and falling limbs of the inflow hydrograph and is difficult to employ to multi-peak hydrographs [42]. There are also issues related to applying the model to the forecasting phase, as discussed later in Section 7.7.1.

In 2015, based on the evaluation of 15 special cases of the model of Eq. (7.29), Easa [40] proposed the following six-parameter model with lateral flow

$$S = K\{w(u)[(1 + \varphi)\,I]^\alpha + (1 - w(u))Q^\alpha\}^\beta, \quad \text{(6P-V-L)} \quad (7.36)$$

In this model, there are four constant parameters $(K, \varphi, \alpha, \beta)$ and one variable parameter $w(u)$ which is considered to have two inflow sub-regions.

In 2015, Karahan et al. [43] implemented the lateral flow parameter in the 3P-C2 model of Eq. (7.13), resulting in a 4P-C-L model, as follows

$$S = K[w(1 + \varphi)I + (1 - w)Q]^\beta, \quad \text{(4P-C-L)} \quad (7.37)$$

The cuckoo search algorithm was used in the calibration of model parameters.

In 2017, Ayvaz and Gurarslan [44] proposed a new partitioning approach for calibrating the 3P-C2 model with variable parameters and lateral flow, resulting in the following model

$$S = K(u)[w(u)(1 + \varphi)\,I + (1 - w(u))Q]^{\beta(u)}, \quad [(1 + 3NL)\text{P-V-L}] \quad (7.38)$$

The inflow hydrograph was partitioned into a finite number of sub-regions, each had its own variable parameters (Fig. 7.4). The number of model parameters equals $(1 + 3NL)$. The model also included one (defining) parameter for each sub-region. The proposed approach was integrated into an optimization model solved using differential evolution. To determine the optimal number of sub-regions, a termination criterion based on improvement rate was used. This interesting partitioning approach is general and can be used with different flood routing procedures.

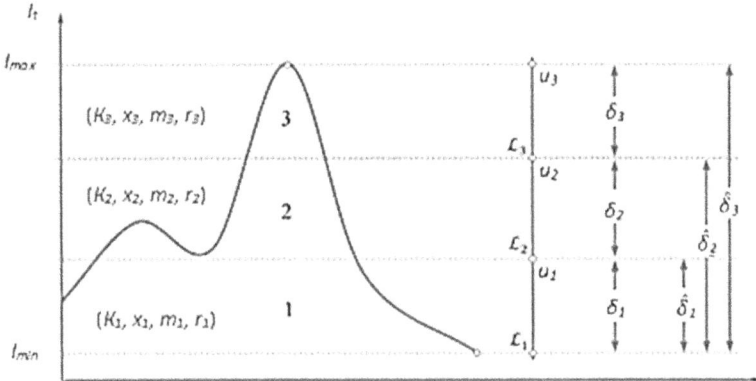

Fig. 7.4. A representative inflow hydrograph which is partitioned into three sub-regions [44].

In 2017, Zhang et al. [45] integrated the lateral flow of Eq. (7.14) by O'Donnell and the variable-exponent parameter model by Easa [17] into a new nonlinear Muskingum model that was solved using an adaptive genetic algorithm. The model is given by

$$S = K[w(1 + \varphi) \, I + (1 - w)Q]^{\beta(u)}, \quad \text{(5P-V-L)} \qquad (7.39)$$

In 2017, Kang et al. [46] incorporated the lateral flow parameter into the two Muskingum models proposed by Easa [18, 3]: the 4P-C2 model of Eq. (7.20) and the 5P-V model of Eq. (7.28). The evaluation of the two models with lateral flow showed that the model with a variable exponent parameter, Eq. (7.28), substantially reduced SSQ. This finding highlights the significance of the variable exponent parameter and is consistent with the results of Easa [3].

7.5.5. Models with Multiple Criteria

Traditional Muskingum models adopt a single criterion in the calibration process. Some models minimize the sum of the squared deviations between estimated and observed outflows (outflow criterion), while others minimize the sum of the squared deviations between the estimated and observed storages (storage criterion). However, models that use the outflow criterion produce large deviations between actual and estimated storages as they do not include a storage constraint. On the other hand, models that use the storage criterion result in a poor fit to the observed outflows.

The storage criterion has been previously formulated by some researchers, including Yoon and Padmanabhan [48] who minimized the sum of the squared deviations between actual and estimated storages. Das [49] took the calibration process one step further by combining the routing equations and the storage model to determine the best fit to observed outflows.

In 2015, Easa [47] developed a new approach that incorporated both criteria in the calibration process. This multi-criteria approach aids trade-off analysis and provides the designer with flexibility in considering the relative merits of these criteria. The multi-criteria function, Z, is defined as

$$Z = \lambda P_Q + (1 - \lambda)P_R, \tag{7.40}$$

where λ is the weight of the normalized outflow criterion specified by the user (0 to 1), P_Q is the normalized outflow criterion (percent deviation from the ideal outflow criteria), and P_R is the normalized storage criterion (percent deviation from the ideal storage criteria). The normalized criteria are used in the multi-criteria function instead of SSQ and SSR since the units of these criteria are substantially different. The storage and outflow criteria are given by

$$SSR = \sum_{j=0}^{N} (\bar{S}_j - S_j)^2 \tag{7.41}$$

$$SSQ = \sum_{j=0}^{N} (\bar{Q}_j - Q_j)^2, \tag{7.42}$$

where $\bar{S}_j$, S_j are the estimated and observed storages for time interval j, respectively, and $\bar{Q}_j$, Q_j are the estimated and observed outflows for time interval j, respectively.

In 2016, Luo et al. [50] developed a multi-criteria calibration of the 3P-C2 model of Eq. (7.13) using non-dominated sorting genetic algorithm (NSGA-II). This multi-criteria software uses evolutionary computation to obtain a set of non-dominated solutions. The authors considered two criteria in model calibration: (a) minimization of SSQ, given by Eq. (7.42) and (b) minimization of the absolute deviations between the peaks of estimated and observed outflows (DPO). The DPO criteria is given by

$$DPO = \sum_{i=1}^{M} \left| Q_i^{Peak\ Obs} - Q_i^{PeakEst} \right|, \qquad (7.43)$$

where $Q_i^{Peak\ Obs}, Q_i^{Peak\ Est}$ are the observed and estimated maximum outflow at peak flow event i, respectively, and M is the number of peak flow events. The software generates a Pareto front for SSQ and DPO, as shown in Fig. 7.5.

The solutions along the Pareto front are optimal and the curve provides a trade-off between the two criteria. Thus, by moving from A1 to C1 along the Pareto front, a reduction of the deviations of the peaks of routed and actual outflows is obtained at the expense of an increase in SSQ. The user would select one specific solution based on the relative importance of the two criteria. The Pareto front exhibits a nonlinear portion within which DPO rapidly decreases, compared with the increase in SSQ. An example of a trade-off solution is shown as point B1 in Fig. 7.5.

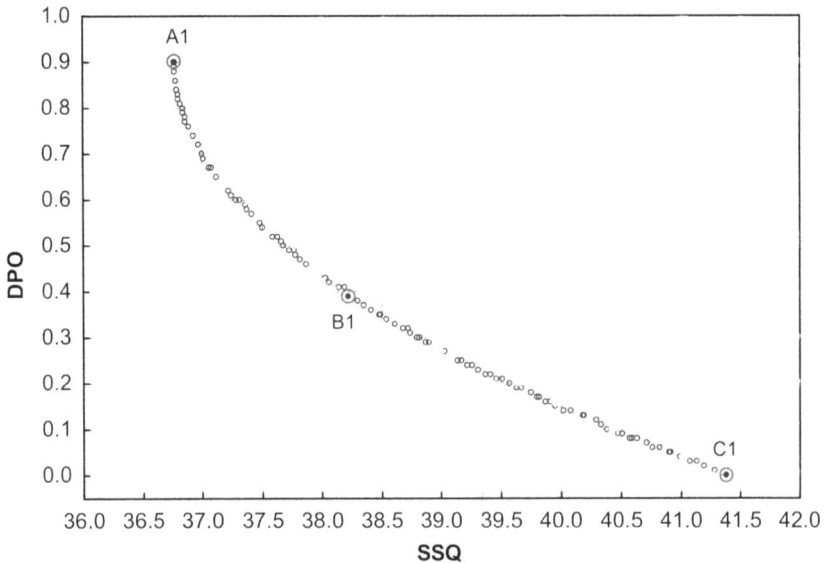

Fig. 7.5. Pareto front for single-peak outflow hydrograph [50].

7.6. Routing Procedures

Two types of routing procedures have been used in the literature for flood routing using nonlinear Muskingum models: modified Euler and fourth-order Runge-Kutta. These procedures are briefly outlined here to provide the reader with a background for subsequent sections.

7.6.1. Modified Euler

The modified Euler (ME) routing procedure, based on Tung [6], is illustrated using the 3P-C2 nonlinear Muskingum model of Eq. (7.13). The routing procedures involve the following steps:

Step 1. Assume values of the three hydrologic parameters, K, w, and β.

Step 2. Calculate the initial storage using Eq. (7.13), where the initial outflow is assumed to be equal to initial inflow ($Q_o = I_o$).

Step 3. Calculate the time rate of change of the storage volume as

$$\frac{\Delta S_j}{\Delta t} = -\left(\frac{1}{1-w}\right)\left(\frac{S_j}{K}\right)^{\frac{1}{\beta}} + \left(\frac{1}{1-w}\right)I_j, \qquad (7.44)$$

where $\Delta S_j / \Delta t$ is the rate of change of the storage volume with respect to time, ΔS_j is the change in the storage volume, and Δt is the time interval.

Step 4. Calculate the next accumulated storage as

$$S_{j+1} = S_j + \Delta S_t \qquad (7.45)$$

Step 5. Calculate the next outflow for $(j + 1)$ as

$$Q_j = \left(\frac{1}{1-w}\right)\left(\frac{S_j}{K}\right)^{\frac{1}{\beta}} - \left(\frac{w}{1-w}\right)I_j \qquad (7.46)$$

Note that in the literature Q_j has been considered as a function of I_j, I_{j-1}, average of I_j and I_{j-1}, or as a function of a weighted average of the two inflows $[\theta I_j + (1 - \theta) I_{j-1}]$, where θ is a parameter [31].

Step 6. Repeat Steps 3 to 5 for the following time intervals.

7.6.2. Fourth-Order Runge-Kutta

Runge–Kutta methods are a family of numerical iterative methods, developed around 1900 by the German mathematicians C. Runge and M. Kutta. The most widely known member of this family is the fourth-order Runge-Kutta (FORK) which has been used in many engineering applications. In its application to flood routing, the FORK procedure can be viewed as a micro-simulation of the storage rate of each time interval, where four storage rates are simulated within the interval and the average storage rate is used to calculate the accumulated storage for the next time interval. Such a scheme would be useful especially for hydrographs that exhibit sudden peaks. The procedure is briefly outlined here for the special case ($NL = 2$) of the model of Eq. (7.16), as an example [17].

Let the number of time intervals of the flood be denoted by N. The inflow, outflow, and weighted storage for time interval j are denoted by I_j, Q_j, and S_j, respectively, where $j = 0, 1, ..., N$. The FORK procedure involves the following steps:

Step 1. Assume values of the three parameters K, w, and $\beta(u_j)$.

Step 2. Calculate the accumulated storage for $j = 0$ as

$$S_j = K[wI_j + (1-w)Q_j]^{\beta(u_j)}, \tag{7.47}$$

where K is the storage parameter, w is the weighting parameter, and β is the variable exponent parameter.

Step 3. Calculate the accumulated storage for $j = 1, 2, ..., N$ as

$$S_j = S_{j-1} + \frac{\Delta t}{6}(k1_{j-1} + 2k2_{j-1} + 2k3_{j-1} + k4_{j-1}), \tag{7.48}$$

where the second term of the right side of Eq. (7.48) is the product of the size of the time interval Δt and an estimated average storage rate based on the storage rates within the interval ($k1_{j-1}$, $k2_{j-1}$, $k3_{j-1}$, and $k4_{j-1}$) which are given by

$$k1_{j-1} = -\left(\frac{1}{1-w}\right)\left(\frac{S_{j-1}}{K}\right)^{\frac{1}{\beta(u_{j-1})}} + \left(\frac{1}{1-w}\right)I_{j-1}, \tag{7.49}$$

$$k2_{j-1} = -\left(\frac{1}{1-w}\right)\left(\frac{S_{j-1}+0.5k1_{j-1}\Delta t}{K}\right)^{\frac{1}{\beta(u_{j-0.5})}} + \left(\frac{1}{1-w}\right)\left(\frac{I_{j-1}+I_j}{2}\right), \quad (7.50)$$

$$k3_{j-1} = -\left(\frac{1}{1-w}\right)\left(\frac{S_{j-1}+0.5k2_{j-1}\Delta t}{K}\right)^{\frac{1}{\beta(u_{j-0.5})}} + \left(\frac{1}{1-w}\right)\left(\frac{I_{j-1}+I_j}{2}\right), \quad (7.51)$$

$$k4_{j-1} = -\left(\frac{1}{1-w}\right)\left(\frac{S_{j-1}+k3_{j-1}\Delta t}{K}\right)^{\frac{1}{\beta(u_j)}} + \left(\frac{1}{1-w}\right)I_j. \quad (7.52)$$

Step 4. Calculate the outflow for $j = 0, 1, \ldots, N$ as

$$\overline{Q}_j = \left(\frac{1}{1-w}\right)\left(\frac{S_j}{K}\right)^{\frac{1}{\beta(u_j)}} - \left(\frac{w}{1-w}\right)I_j \quad (7.53)$$

Note that the calculated outflow for $j = 0$ always equals the observed outflow Q_0.

7.7. Practical Considerations

7.7.1. Guidelines for Model Calibration

In calibrating Muskingum models, one should realize that the calibrated model will be subsequently used for forecasting, as previously illustrated in Fig. 7.1. However, significantly more attention is usually given by researchers to model calibration than forecasting. The importance of sound model calibration that is reliable for forecasting is stressed [52]. As pointed out by Bair [53], a hydrologic expert witness will usually use model calibration as a critical part of the scientific methods applied to the construction of a forecasting model. In this regard, the following important issues are highlighted:

(a) The calibrated model should not have many parameters as this would reduce the reliability of the model in the forecasting phase. One frequent hydrologic problem in model calibration is that one has a model in which some critical parameters are modeled in such a way that it estimates the system output quite accurately, yet its forecasting ability is not good [52]. Although models with more than

10 parameters have been proposed in the literature, the guidelines for model selection presented later recommend models with up to six parameters to be used for different types of hydrographs (SPS, SPN, and MP).

(b) For a model to be useful in forecasting, the parameters need to accurately reflect the properties of the components of the underlying system, as stressed by Yilmaz et el. [54]. This is particularly relevant when developing a function for the variable parameters of the Muskingum model. The variable parameter should be a function of a flood characteristic, not merely a function of time. An example of such a function is the inflow-based discontinuous parameter used in Easa [17]. On the other hand, an example of a function that is flood irrelevant is Eq. (7.31) which is a function of only time.

(c) A common problem in the calibration phase is that the model is calibrated using flood-specific characteristics that are different from those of the forecasted flood. For example, consider Eq. (7.31) that expresses the exponent parameter as a continuous function of t, where t ranges from 0 to T (flood duration). Then, the calibrated continuous parameter is valid for only a specific flood duration. Since flood events in the same or similar river reaches do not usually occur with the same duration, application of such a continuous parameter would be questionable if the duration of the forecasted flood is different from that used in calibration. In particular, extrapolation is dangerous if future flood duration is much larger than T as model performance may deteriorate rapidly [52].

(d) Another example of using a specific flood characteristic is the peak time of the inflow hydrograph, t_p, used in the model by Karahan [31], where two sets of parameters corresponding to the period before and after the peak are estimated. Thus, if the peak times in calibration and forecasting are substantially different, the calibrated parameters would be unreliable. This problem is complicated by the fact that the lateral flow affects the peak of the hydrograph, as stated by Wang [55], and this effect is likely to be different for the inflow hydrographs used in the calibration and forecasting phases. In addition, it is known that for river reaches with significant lateral flow, the inflow hydrograph has multiple peaks which would make t_p meaningless. To avoid the problems of using flood-specific characteristics, dimensionless entities should be used in calibration. The inflow-based dimensionless function suggested by Easa [17] provided a good basis for developing similar or improved functions by other researchers, such as Vatankhah [32], Afzali [41], Niazkar and Afzali [42], Ayvaz and Gurarslan [44], Zhang et al. [45], and Kang et al. [46].

(e) Finally, one should be cautious when applying Muskingum models with lateral flow parameters. If the lateral flow does not actually exist or is negligible in the river reach, the lateral flow parameter would be an extra meaningless parameter that just helps to improve model fit, making the estimated parameters unreliable. To illustrate this issue, using Wilson's data, O'Donnell [23] estimated the three-parameter model and obtained a positive value of the lateral flow parameter (lateral inflow). However, for the same data Karahan [31] estimated the 10-parameter model and obtained a negative value of the parameter (lateral outflow). It is, therefore, necessary to examine carefully catchment drainage behavior and spatial storm distribution patterns to better understand lateral flow characteristics before deciding to use a Muskingum model with lateral flow [3]. This examination helps not only to decide whether a lateral flow parameter is needed, but also to determine the type of lateral flow.

7.7.2. Continuous vs. Discontinuous Parameters

A comparison of the accuracy of ME and FROK procedures using continuous and discontinuous variable exponent parameters has been undertaken by Easa [51] in 2015. Two functional forms of the continuous parameter (logarithmic and exponential) combined with two different routing procedures (ME and FORK) were evaluated. The parameter was expressed as a function of a dimensionless inflow variable. The results showed that the trend of the optimal values of the parameters resembles that of the inflow. It appears that the calibrated optimal parameter carries the signature of the inflow. This may provide support for using the inflow as a relevant flood property for defining the exponent parameter and its transferability to the forecasting phase. In addition, the model with continuous parameter was straightforward to calibrate and faster to run. For multi-peak hydrographs, the discontinuous parameter provided substantially better performance and is recommended for such hydrographs. For smooth and non-smooth hydrographs, the continuous parameter provided better performance when combined with the ME procedure than when combined with the FORK procedure. However, the

FORK procedure may perform better when another criteria such as storage is used [47].

The performance of the ME and FORK procedures were compared using the three types of hydrographs [3]. The continuous outperformed the discontinuous parameter for the SPS and SPN hydrographs only. For the MP hydrograph, the discontinuous parameter was better, as shown in Table 7.5. The discontinuous parameter substantially improved model performance using ME and FORK. Moreover, the continuous and discontinuous parameters appear to aid FORK performance. This is not surprising as FORK can be viewed as a micro-simulation of the storage rate within each time interval [56], where four storage rates are simulated and the average storage rate is used to calculate the accumulated storage for the next time interval.

Table 7.5. Performance of the routing procedures for MP hydrograph.

Exponent Parameter	SSQ for ME	SSQ for FORK
Constant β	71,114	68,189
Continuous β	64,792	63,300 [b]
Discontinuous β [a]	44,894	28,268
Diff. (cont. and discount.) (%)	30.7	55.3

[a] Based on $NL = 5$. [b] Based on the continuous function, $m(uj) = a + b \ln(1 + c\ uj)$.

7.7.3. Guidelines for Selection of Model Type and Routing Procedure

It is clear from the previous sections that a large number of Muskingum models have been developed in recent years. These developments and the availability of different routing procedures have raised the need to establish preliminary guidelines for the selection of model type and routing procedure for various types of hydrographs.

For the selection of model type, the 15 Muskingum models with different combinations of constant and variable parameters (Table 7.4) were evaluated for three types of hydrographs by Easa [40]. Based on this evaluation, guidelines for the selection of model type were established (Table 7.6). The recommended models have four to six parameters and substantially improve model performance compared with the traditional 3P-C2 model. For SPS hydrographs, the recommended model is the 4P-C2 model. For SPN hydrographs, the recommended model is 5P-V, where w is a variable parameter. For MP hydrographs, the recommended model is 6P-V, where w and β are variable parameters. The variable parameters were based on two-inflow sub-regions.

The recommended models were selected based a trade-off between the marginal reduction in SSQ and the corresponding number of parameters. This is illustrated in Fig. 7.6 for the SPS hydrograph. The figure shows the relation between SSQ and the number of model parameters. As noted, beyond four parameters, the SSQ reduction is small. Therefore, the 4P-C2 model (4P-MM in the figure) is recommended for this hydrograph type.

For the selection of routing procedure, guidelines have been presented for different types of hydrographs [40], as shown in Table 7.7. For each hydrograph type and exponent parameter type (continuous and discontinuous), the recommended routing procedure (ME and FORK) is shown. For the SPS hydrograph, the best results are obtained with continuous exponent parameter and ME. For the SPN hydrograph, the best results are obtained with continuous exponent parameter and ME or FORK. For the MP hydrograph, the best results are obtained with discontinuous exponent parameter and FORK.

Note that the guidelines for selection of routing procedure were established based on the evaluation of only three hydrographs. Further studies to evaluate a larger number of hydrographs with different trends are obviously needed to verify or modify these guidelines.

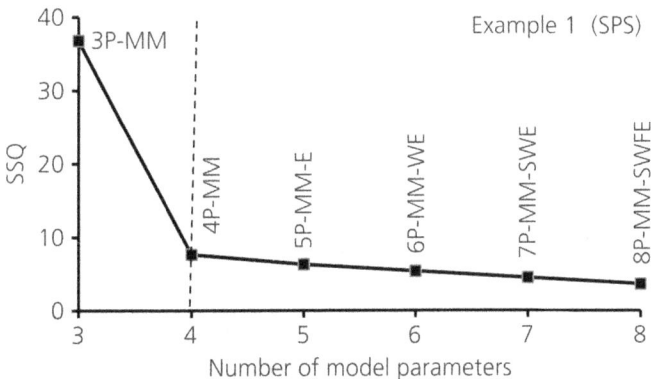

Fig. 7.6. Variation of model performance with number of model parameters (SPS hydrograph) [40].

283

Table 7.6. Guidelines for selection of model type.

Hydro. Type	Model Designation	Storage Equation [a]
SPS	4P-C2	$S = K[wI^\alpha + (1-w)Q^\alpha]^\beta$
SPN	5P-V (variable w)	$S = K[\mathrm{w(u)}I^\alpha + (1-w(\mathrm{u}))Q^\alpha]^\beta$
DP	6P-V (variable w and β)	$S = K[\mathrm{w(u)}I^\alpha + (1-w(\mathrm{u}))Q^\alpha]^{\beta(\mathrm{u})}$

[a] $w(u)$ and $\beta(u)$ are variable parameters represented by two-step functions.

Table 7.7. Guidelines for selection of routing procedure.

Hydro. Type	Variable Exponent Type	
	Continuous	Discontinuous
SPS	ME	–
SPN	ME or FORK	–
MP	–	FORK

7.7.4. Plagiarism and Ethical Issues

Plagiarism and ethical issues may arise after new ideas are published. Given this serious matter, let us first highlight the following statement by Plagramme [57] *"Plagiarism is an act when you take someone's work and try to pass it off as your own. This is known as stealing, which is not only unethical, but it is also illegal. In legal terms, plagiarism is considered literary theft. If you knowingly use another person's work without giving them credit, you are committing plagiarism."* Regarding the new four-parameter (4P) model which was introduced by Easa [18] in 2014 as part of the alternative thinking, there were four cases (from two universities) that seriously violated research ethics.

In the first case, after the new 4P-model paper was published ahead of print in 2014, Vatankhah [36] presented in a discussion (of another paper) a 5P model, adopting the alternative thinking proposed by Easa [18], without citing the 4P model. As if this was not enough, the author included numerous statements that were taken almost word by word from Easa [18] to justify his 5P model. This case was discussed in detail in a closure by Easa [3] in 2014. In 2017, the author admitted that the 4P model was the first effort to modify the structure of the 3P model [32].

In the second case, in 2016 Hamedi et al. [58] referred to the new 4P model by Easa [18] as their own model! In the Introduction, the authors state *"The third nonlinear Muskingum model (NL3) was introduced by*

Hamedi et al. (2014a, b) using Eq. (5)." In their paper, Eq. (5) is identical to the 4P model of Eq. (7.20) by Easa [18] and the preceding references refer to two discussions by Hamedi et al. [59, 60] that did not include any mention of a four-parameter model. Clearly, the authors falsely took credit for the 4P model. Ironically, the preceding statement was written (differently) in the proof of the same paper (posted on ResearchGate as a public document) as *"The third nonlinear Muskingum model (NL3) was introduced by Jesus (2013) using Eq. (5)."* The reference Jesus (2013) was not listed in their reference list (Jesus is the English name of the Arabic name Easa). In the final published version, the authors changed the credit for the model to themselves anyway. To make the matter worse, later in the paper the authors referred to Easa [18] as an application of the 4P model by stating *"The NL3 and NL4 parameters were estimated respectively with GA-GRG and SFLA-NMS by Easa (2013) and Bozorg-Haddad et al. (2015c)."* The reference Easa (2013) is the 4P model itself that they took credit for, not an application of the model (date should be 2014). The authors' action is incredibly troubling.

In the third case, even though the 4P model was the *core* of their study, in 2018 Farahani et al. [61] did not cite the 4P model by Easa [18] in their paper and implicitly gave the credit to another author. In the Introduction section, the authors cited two dozen references and then stated *"the 4-parameter Muskingum model is of a higher precision than the 2-parameter and 3-parameter Muskingum models, so that the 4-parameter model is used as a high-performance model (Barati 2018)."* Clearly, in the absence of citation of the 4P model, this statement misrepresents the credit for the model. Subsequently, in the History section, the authors described several papers (that had implemented the 4P model) and incorrectly referred to the 4P model by citing Easa (2015). This reference is an application of the model, Easa [51], not the model itself, which was published in 2014. The authors' misrepresentation of the 4P model, which was the focus of their study, is reprehensible.

In the fourth case, in 2018 Farzin et al. [62] misrepresented the development and impact of the 4P model by stating *"Four parameters were considered as decision variables, and the results indicated that the model based on the considered parameters with the genetic algorithm decreased the RMSE and mean absolute error (MAE) by 20 % and 25 %, respectively, compared to the nonlinear programming methods [24]."* Reference [24] is the 4P model by Easa [18]. Beside the inconsistency in citation, the preceding statement is erroneous in three respects: (a) the statement does not explicitly mention the development of the 4P model;

a model that has substantially improved the fit to observed outflows after unsuccessful long-time effort of modifying the solution algorithms of the 3P model, (b) the 20 % improvement by the 4P model mentioned in the statement is incorrect as the improvement is up to 80 %, and (c) the improvement by the 4P model is not in comparison with *"nonlinear programming methods",* but with the traditional 3P model.

It is hoped that the exposure of the preceding cases will be beneficial to researchers. The intent is to discourage some researchers (admittedly few) from plagiarism and unethical practices. Those should realize that the world now is a global village with advanced technologies that can easily expose research misconducts.

7.8. Concluding Remarks

The alternative thinking of improving the performance of the traditional three-parameter Muskingum model by modifying parameter and model structures [17, 18], rather than solution algorithms, has produced great results. During the past five years, at least 34 new hydrologic Muskingum models with four or more (constant and variable) parameters have been developed. These new models have provided better performance than the traditional 3P model that has been used in hydrology for 35 years. The chapter has also reviewed the main flood routing procedures and discussed several practical considerations regarding model calibration, use of continuous and discontinuous parameters, selection of model type and routing procedure, and plagiarism and ethical issues.

The most important message of this chapter is that the focus on improving model performance during calibration should be accompanied by equal focus on the implications of the calibrated model in forecasting. Such implications should be thoroughly examined as the process may become a legal issue.

Despite the surge of research work on hydrologic Muskingum models during the past five years, the following opportunities for further development can be explored:

1. In parallel to the alternative thinking of modifying parameter and model structures, research should continue to develop better solution algorithms. For example, Yuan et al. [63] recently proposed an

improved backtracking search algorithm (BSA) for estimating the parameters of the nonlinear Muskingum model. The authors introduced an orthogonal designed initialization population strategy and chaotic sequences to improve the exploration and exploitation ability of BSA. The results showed that the proposed BSA outperformed PSO, genetic algorithm, and differential evolution.

2. The practice of uncertainty analysis in hydrologic engineering is not widespread. Limited research on probabilistic Muskingum models has been conducted, including Das [64] and Ahmadisharaf et al. [65]. More research is needed to incorporate uncertainty in flood management and decision-making. Excellent references that promote the uptake of uncertainty analysis methods to improve flood risk management include Beven and Hall [66], Hall [67], and Liu and Gupta [68].

3. Multi-criteria optimization of the model has been emerging and better solution algorithms can be developed for model calibration. In particular, fuzzy multi-criteria can be explored for calibration of Muskingum models. Another strategy to improve forecasting accuracy is dynamic parameter estimation. For example, Zhang et al. [69] has proposed an "in-process type" method using back-propagation neural network model for estimating highly-accurate Muskingum models.

4. Previous studies have focused on producing better calibrated models. No comprehensive study, to the writer's knowledge, has been conducted to verify the calibrated models using independent flood events. Thus, more research and case studies are needed to evaluate the reliability of the calibrated models for use in forecasting.

Acknowledgements

Writing this chapter has been a fantastic adventure. I wish to thank all researchers who have contributed to the recent developments of the hydrologic Muskingum models presented in this chapter. While reading various papers, I have enjoyed their fine thoughts and contributions. I wish also to give a valuable advice to researchers, especially young ones. Publication of new ideas is sometimes challenging and may be delayed because of resistance to change or other reasons. In fact, the two papers [17, 18] that have initiated the alternative thinking of modifying model and parameter structures instead of solution algorithms were originally declined, but later published after persistence. If you have a new idea, I encourage you to persevere in publishing it. There are many wonderful people around who will appreciate your work and help you to grow.

References

[1]. K. Subramanya, Engineering Hydrology, *McGraw Hill*, New York. N.Y., 2008.

[2]. G. T. McCarthy, The unit hydrograph and flood routing, in *Proc., Conference of North Atlantic Division*, U. S. Army Corps of Engineers, Rhode Island, 1938.

[3]. S. M. Easa, Closure to "Improved nonlinear Muskingum model with variable exponent parameter," *J. Hydrol. Eng.*, Vol. 19, Issue 10, 2014, pp. 1-10.

[4]. V. T. Chow, Open Channel Hydraulics, *McGraw Hill*, New York, N.Y., 1959.

[5]. M. A. Gill, Flood routing by the Muskingum method, *J. Hydrol.*, Vol. 36, 1978, pp. 353-363.

[6]. Y. K. Tung, River flood routing by nonlinear Muskingum method, *J. Hydraul. Eng.*, Vol. 111, Issue 12, 1985, pp. 1147-1460.

[7]. S. Mohan, Parameter estimation of nonlinear Muskingum models using genetic algorithm, *J. Hydraul. Eng.*, Vol. 123, Issue 2, 1997, pp. 137-142.

[8]. J. H. Kim, Z. W. Geem, E. S. Kim, Parameter estimation of the nonlinear Muskingum model using harmony search, *J. American Water Resour. Assoc.*, Vol. 37, Issue 5, 2001, pp. 1131-1138.

[9]. Z. W. Geem, Parameter estimation for the nonlinear Muskingum model using the BFGS technique, *J. Irrig. Drain. Eng.*, Vol. 132, Issue 5, 2006, pp. 474-478.

[10]. H. J. Chu, L. C. Chang, Applying particle swarm optimization to parameter estimation of the nonlinear Muskingum model, *J. Hydrol. Eng.*, Vol. 14, Issue 9, 2009, pp. 1024-1027.

[11]. J. Luo, J. Xie, Parameter estimation for nonlinear Muskingum model based on immune clonal selection algorithm, *J. Hydrol. Eng.*, Vol. 15, Issue 10, 2010, pp. 844-851.

[12]. R. Barati, Parameter estimation of nonlinear Muskingum models using Nelder-Mead simplex algorithm, *J. Hydrol. Eng.*, Vol. 16, Issue 11, 2011, pp. 946-954.

[13]. Z. W. Geem, Parameter estimation of the nonlinear Muskingum model using parameter-setting-free harmony search, *J. Hydrol. Eng.*, Vol. 16, Issue 8, 2011, pp. 684-688.

[14]. D. Xu, L. Qiu, S. Chen, Estimation of nonlinear Muskingum model parameters using differential evolution, *J. Hydrol. Eng.*, Vol. 17, Issue 2, 2012, pp. 348-353.

[15]. H. Orouji, O. B. Haddad, E. Fallah-Mehdipour, M. A. Mariño, Estimation of Muskingum parameter by meta-heuristic algorithm, *ICE Water Manage.,* Vol. 166, Issue 6, 2013, pp. 315-324.

[16]. H. Karahan, G. Gurarslan, Z. W. Geem, Parameter estimation of the nonlinear Muskingum flood routing model using a hybrid harmony search algorithm, *J. Hydrol. Eng.*, 2013, Vol. 18, Issue 3, pp. 352-360.

[17]. S. M. Easa, Improved nonlinear Muskingum model with variable exponent parameter, *J. Hydrol. Eng.*, Vol. 18, Issue 12, 2013, pp. 1790-1794.

[18]. S. M. Easa, New and improved four-parameter nonlinear Muskingum model, *ICE Water Manage.*, Vol. 167, Issue 5, 2014, pp. 288-298.

[19]. E. M. Wilson, Engineering Hydrology, *MacMillan Education Ltd*, Hampshir, U.K., 1974.

[20]. Flood Studies Report, Vol. 3, *Natural Environment Research Council (NERC)*, Wallingford, UK, 1975.

[21]. W. Viessman, G. L. Lewis, Introduction to Hydrology, *Pearson Education Inc.*, Upper Saddle River, New Jersey, 2003.

[22]. G. Gavilan, M. H. Houck, Optimal Muskingum river routing, in *Proc., Computer Applications in Water Resources Specialty Conference,* 1985, pp. 1294-1302.

[23]. T. O'Donnell, A direct three-parameter Muskingum procedure incorporating lateral inflow, *J. Hydrol. Sci.,* Vol. 30, Issue 4, 1985, pp. 479-496.

[24]. T. O'Donnell, Improved fitting for three-parameter Muskingum procedure. *J. Hydraul. Eng.*, Vol. 114, Issue 5, 1988, pp. 516–528.

[25]. H. M. Khan, Muskingum flood routing model for multiple tributaries. *Water Resour. Res.*, Vol. 29, Issue 4, 1993, pp. 1057–1062.

[26]. P. Choudhury, R. K. Shrivastava, S. M. Narulkar, Flood routing in river networks using equivalent Muskingum inflow, *J. Hydrol. Eng.*, Vol. 7, Issue 6, 2002, pp. 413-419.

[27]. A. Azadnia, B. Zahraie, Optimization of nonlinear Muskingum method with variable parameters using multi-objective particle swarm optimization, in *Proc., World Environmental & Water Resources Congress,* Providence, Rhode Island, 2010, pp. 2278-2284.

[28]. A. R. Vatankhah, Discussion of 'Improved nonlinear Muskingum model with variable exponent parameter, by S. M. Easa, Vol. 18, Issue 12, 2013, pp. 1790-1794', *J. Hydrol. Eng.*, Vol. 19, Issue 10, 2014, pp. 1-4.

[29]. F. Hamedi, O. B. Haddad, H. Orouji, Discussion of 'Improved nonlinear Muskingum model with variable exponent parameter, by S. M. Easa, Vol. 18, Issue 12, 2013, pp. 1790-1794', *J. Hydrol. Eng.*, Vol. 19, Issue 10, 2014, pp. 1-3.

[30]. J. Luo, X. He, Discussion of 'Improved nonlinear Muskingum model with variable exponent parameter, by S. M. Easa, Vol. 18, Issue 12, 2013, pp. 1790-1794', *J. Hydrol. Eng.*, Vol. 19, Issue 10, 2014, pp. 1-4.

[31]. H. Karahan, Discussion of 'Improved nonlinear Muskingum model with variable exponent parameter, by S. M. Easa, Vol. 18, Issue 12, 2013, pp. 1790-1794', *J. Hydrol. Eng.*, Vol. 19, Issue 10, 2014, pp. 1-9.

[32]. A. R. Vatankhah, Non-linear Muskingum model with inflow-based exponent, *ICE Water Manage.*, Vol. 170, Issue 2, 2017, pp. 66-80.

[33]. A. Moghaddam, J. Behmanesh, A. Farsijani, Parameters estimation for the new four-parameter nonlinear Muskingum model using the particle-swarm optimization, *Water Resour. Manage.*, Vol. 30, 2016, pp. 2143–2160.

[34]. H. Karahan, Closure to 'Parameter Estimation of the Nonlinear Muskingum Flood-Routing Model Using a Hybrid Harmony Search Algorithm, by H. Karahan, G. Gurarslan, and Z. Woo Geem, Vol. 18, Issue 3, 2013, pp. 352–360', *J. Hydrol. Eng.*, Vol. 19, Issue 4, 2014, pp. 847–853.

[35]. S. M. Easa, R. Barati, H. Shahheydari, E. Nodoshan, T. Barati, Discussion: 'New and improved four-parameter non-linear Muskingum flood model, by S. M. Easa, Vol. 167, Issue 5, 2014, pp. 288-298', *ICE Water Manage.*, Vol. 167, Issue 10, 2014, pp. 612-615. Note: Discussion by R. Barati et al. and Reply by S. M. Easa.

[36]. A. Vatankhah, Discussion of 'Parameter estimation of the nonlinear Muskingum flood-routing model using a hybrid harmony search algorithm, by H. Karahan, G. Gurarslan, Z. Geem, 2013, Vol. 18, Issue 3, pp. 352-360', *J. Hydrol. Eng.*, Vol. 19, Issue 4, 2014, pp. 839-842.

[37]. O. B. Haddad, F. Hamedi, H. Orouji, M. Pazoki, H. A. Loáiciga, A re-parameterized and improved nonlinear Muskingum model for flood routing. *Water Resour. Manage.*, Vol. 29, Issue 9, 2015, pp. 3419-3440.

[38]. M. Niazkar, S. H. Afzali, New nonlinear variable-parameter Muskingum models, *KSCE J. Civ. Eng.*, Vol. 21, Issue 7, 2017, pp. 2958-2967.

[39]. O. B. Haddad, F. Hamedi, E. Fallah-Mehdipour, Application of a hybrid optimization method in Muskingum parameter estimation, *J. Irrig. Drain. Eng.*, Vol. 141, Issue 12, 2015, 04015026.

[40]. S. M. Easa, Versatile Muskingum flood model with four variable parameters, *ICE Water Manage.*, Vol. 168, Issue 3, 2015, pp. 139-148.

[41]. S. H. Afzali, Variable-parameter Muskingum model, *Iranian J. Science and Technology, Transactions of Civil Eng.*, Vol. 40, Issue 1, 2016, pp. 59-68.

[42]. M. Niazkar, S. H. Afzali, Application of new hybrid optimization technique for parameter estimation of new improved version of Muskingum model, *Water Resour. Manage.*, Vol. 30, 2016, pp. 4713-4730.

[43]. H. Karahan, G. Gurarslan, Z. W. Geem, A new nonlinear Muskingum flood routing model incorporating lateral flow, *Eng. Optim.*, Vol. 47, 2015, pp. 737-749.

[44]. M. T. Ayvaz, G. Gurarslan, A new partitioning approach for nonlinear Muskingum flood routing models with lateral flow contribution, *J. Hydrol.*, Vol. 553, 2017, pp. 142-159.

[45]. S. Zhang, L. Kang, L. Zhou, X. Guo, A new modified nonlinear Muskingum model and its parameter estimation using the adaptive genetic algorithm, *Hydrol. Res.*, Vol. 48, Issue 1, 2017, pp. 17-27.

[46]. L. Kang, L. Zhou, S. Zhang, Parameter estimation of two improved nonlinear Muskingum models considering the lateral flow using a hybrid algorithm, *Water Resour. Manage.*, Vol. 31, 2017, pp. 4449-4467.

[47]. S. M. Easa, Multi-criteria optimization of Muskingum flood model: A new approach, *ICE Water Manage.*, Vol. 168, Issue 5, 2015, pp. 220-231.

[48]. J. Yoon, G. Padmanabhan, Parameter estimation of linear and nonlinear Muskingum models, *J. Water Resour. Plann. Manage.*, Vol. 119, Issue 5, 1993, pp. 600-610.

[49]. A. Das, Parameter estimation for Muskingum models, *J. Irrig. Drain. Eng.*, Vol. 130, Issue 2, 2004, pp. 140-147.

[50]. J. Luo, X. Zhang, X. Zhang, Multi-objective calibration of nonlinear Muskingum model using non-dominated sorting genetic algorithm-II, in *Proc., International Conference on Applied Mathematics, Simulation and Modelling*, Amsterdam, The Netherland, 2016.

[51]. S. M. Easa, Evaluation of nonlinear Muskingum model with continuous and discontinuous exponent parameters, *KSCE J. Civ. Eng.*, Vol. 19, Issue 7, 2015, pp. 2281-2290.

[52]. A. Morton, M. Suarez, Kinds of models, Chapter 2, in Model Validation: Perspective in Hydrological Science (M. G. Anderson, P. D. Bates, Eds.), *John Wiley*, New York, 2001.

[53]. E. S. Bair, Models in the courtroom, Chapter 5, in Model Validation: Perspective in Hydrological Science (M. G. Anderson, P. D. Bates, Eds.), *John Wiley*, New York, 2001.

[54]. K. K. Yilmaz, et al., Model calibration in watershed hydrology, Chapter 3, Advances in Data-Based Approaches for Hydrological Modeling and Forecasting (B. Sivakumar, R. Berndtsson, Eds.), *World Scientific*, New Jersey, N.J., 2010.

[55]. B. E. Wang, F. Q. Tian, H. P. Hu, Analysis of the effect of regional lateral inflow on the flood peak of Three Gorges Reservoir, *China Technological Sciences*, Vol. 54, Issue 4, 2011, pp. 914-923.

[56]. C. F. Gerald, P. O. Wheatley, Applied Numerical Analysis, *Addison-Wesley*, Boston, M.A., 1994.

[57]. Plagramme Company, Ethics of plagiarism, Internet: https://www.plagramme.com/ethics-of-plagiarism (accessed September 21, 2018).

[58]. F. Hamedi, O. Bozorg-Haddad, M. Pazoki, H-R. Asgari, M. Parsa, H. A. Loáiciga, Parameter estimation of extended nonlinear Muskingum models with the weed optimization algorithm, *J. Irrig. Drain. Eng.*, Vol. 142, Issue 12, 2016, pp. 1-11.

[59]. F. Hamedi, O. Bozorg-Haddad, H. Orouji, E. Fallah-Mehdipour, M. A. Marino, Discussion of 'Parameter estimation of the nonlinear Muskingum flood-routing model using a hybrid harmony search algorithm' by H. Karahan, G. Gurarslan, Z. W. Geem, *J. Hydrol. Eng.*, 2014, Vol. 19, Issue 4, 2014, pp. 845 – 847.

[60]. F. Hamedi, O. Bozorg-Haddad, H. Orouji, Discussion of 'Application of excel solver for parameter estimation of the nonlinear Muskingum models' by R. Barati, *KSCE J. Civ. Eng.*, Vol. 19, Issue 1, 2014, pp. 340-342.

[61]. N. N. Farahani, S. Farzin, H. Karami, Flood routing by Kidney algorithm and Muskingum model, *Natural Hazards,* 2018 (paper not assigned to an issue).

[62]. S. Farzin, V. P. Singh, H. Karami, N. Farahani, M. Ehteram, O. Kisi, M. Falah Allawi, N. S. Mohd, A. El-Shafie, Flood routing in river reaches using a three-parameter Muskingum model coupled with an improved bat algorithm, *Water*, Vol. 10, 2018, 1-24.

[63]. X. Yuan, X. Wu, H. Tian, Y. Yuan, R. Adnan, Parameter identification of nonlinear Muskingum model with backtracking search algorithm, *Water Resour. Manage.*, Vol. 30, Issue 8, 2016, pp. 2767-2783.

[64]. A. Das, Chance-constrained optimization-based parameter estimation for Muskingum models, *J. Irrig. Drain. Eng.*, Vol. 133, Issue 5, 2007, pp. 487-494.

[65]. E. Ahmadisharaf, A. Kalyanapu, E-S. Chung, Spatial probabilistic multi-criteria decision making for assessment of flood management alternatives, *J. Hydrol.*, 2016, pp. 1-46.

[66]. K. Beven, J. Hall, Applied uncertainty analysis for flood risk management, *Imperial College Press*, London, England, 2014.

[67]. J. W. Hall, A framework for uncertainty analysis in flood risk management, *Int. J. Basin Manage.*, Vol. 6, 2008, pp. 85-98.

[68]. Y. Liu, H. V. Gupta, Uncertainty in hydrologic modeling: Toward an integrated data assimilation framework, *Water Resour. Res.*, Vol. 43, Issue 3, 2007.

[69]. G. Zhang, T. Xie, L. Zhang, X. Hua, C. Wu, In-process type dynamic Muskingum model parameter estimation method, *Water*, Vol. 9, Issue 11, 2017, pp. 849-869.

Index

E

Eigenvalue, 194
Electrical
    Capacitance Tomography, 154
    Impedance Tomography, 154
    Resistance Tomography, 154
empirical mode decomposition, 85
emulsion, 166
EMYW, 197
envelope, 85, 90, 106
Ergodic process, 201
Error, 142
Errors, 177
    deterministic error, 177
    random error, 177
errors-in-variables model (EIV), 16
Evaluation, 140
Expectation, 181, 200

F

false alarm probability, 111, 113
filter, 103
    band-pass, 107
    ideal, 100
    low-pass, 100
Filtered residual, 189
Final prediction error, 193
Flood
    routing
    modified Euler.method, 277
    Runge-Kutta method, 278
    routing, calibration, 258
    routing, forecasting, 258
    duration, 280
    peak time, 280
Flow
    area, 224
Flow depth
    alternate and sequent, 245
    normal and critical, 243
Flow
    exponents, 269
    uniform, 235
Fourier
    damping, 81
    discrete, 69
    periodic spectrum, 69, 78
    properties of F. transform, 63
Fourier transform, 178

Freeboard, 245
frequency
    instantaneous, 88
    negative, 81, 84
    Nyquist, 77, 84
    shift, 81

G

gamma-ray, 150
gas void fraction, 162
Gevers-Wouters algorithm, 191
Grass-lined channels, 242
Guidelines,
    model type, 282
    routing procedure, 283
GUM uncertainty framework (GUF),
    48, 49
GVF, 162

H

Hardness, 136, 139, 140, 143
Heavy oil, 167
hydrate inhibitors, 167
Hydrograph
    partitioning, 273
Hydrograph,
    inflow, 258
    outflow, 258
    types. *See* single-peak, smooth
    multi-peak, 260
    single-peak, non-smooth, 260

I

Impedance, 153
Inertial sensors, 177
    accelerometer, 177
    gyroscope, 177, 198, 210
Inflow level, 264
integrate, 87

J

jitter, 110, 117

K

Kalman filter, 212